BURLEIGH DODDS SCIENCE: INSTANT INSIGHTS

NUMBER 75

Nutritional and health benefits of beverage crops

Published by Burleigh Dodds Science Publishing Limited
82 High Street, Sawston, Cambridge CB22 3HJ, UK
www.bdspublishing.com

Burleigh Dodds Science Publishing, 1518 Walnut Street, Suite 900, Philadelphia, PA 19102-3406, USA

First published 2023 by Burleigh Dodds Science Publishing Limited

British Library Cataloguing in Publication Data
A catalogue record for this book is available from the British Library

ISBN 978-1-78676-975-6 (Print)
ISBN 978-1-78676-976-3 (ePub)

DOI: 10.19103/9781786769763

Typeset by Deanta Global Publishing Services, Dublin, Ireland

Contents

Series list

Title	Series number
Sweetpotato	01
Fusarium in cereals	02
Vertical farming in horticulture	03
Nutraceuticals in fruit and vegetables	04
Climate change, insect pests and invasive species	05
Metabolic disorders in dairy cattle	06
Mastitis in dairy cattle	07
Heat stress in dairy cattle	08
African swine fever	09
Pesticide residues in agriculture	10
Fruit losses and waste	11
Improving crop nutrient use efficiency	12
Antibiotics in poultry production	13
Bone health in poultry	14
Feather-pecking in poultry	15
Environmental impact of livestock production	16
Pre- and probiotics in pig nutrition	17
Improving piglet welfare	18
Crop biofortification	19
Crop rotations	20
Cover crops	21
Plant growth-promoting rhizobacteria	22
Arbuscular mycorrhizal fungi	23
Nematode pests in agriculture	24
Drought-resistant crops	25
Advances in detecting and forecasting crop pests and diseases	26
Mycotoxin detection and control	27
Mite pests in agriculture	28
Supporting cereal production in sub-Saharan Africa	29
Lameness in dairy cattle	30
Infertility/reproductive disorders in dairy cattle	31
Alternatives to antibiotics in pig production	32
Integrated crop-livestock systems	33
Genetic modification of crops	34

Chapter 1

Beneficial compounds from coffee leaves

Claudine Campa, UMR IPME, France; and Arnaud Petitvallet, Wize Monkey, Canada

1 Introduction

Selection in crop plants is mainly based on performance criteria and product quality. For coffee plants, this selection is essentially based on the production level of coffee beans and on the organoleptic quality of the coffee drink. Many suggestions have been proposed to improve these two criteria, either by modifying cultivation techniques, for example, by input supply, cultivation in full sun or under shade, and plant treatment against biological aggressors or by practicing inter- or intraspecific crosses. However, these improvements only address in part the problems of climate change and development of a sustainable farming.

Climate change may result (Elad and Pertot, 2014), or has already resulted, in a resurgence of pathogen attack, as shown, for example, in some American (Rozo et al., 2012) or Ethiopian *Coffea arabica* plantations (Adugna et al., 2016). New breeding strategies must be found to select performing varieties with enhanced adaptive capabilities, on the basis of their resistance to biotic and abiotic stresses. Adaptation to environmental changes is closely linked to the plant's ability to respond to stress and acclimate. For evergreen plants, the part of the plant most continuously exposed to environmental change – that is, temperature, light or drought, and sensing – is the leaf. This organ, the leading provider of the carbon source for plant growth and reproduction, possesses a wide range of protection systems against environmental stress. Characterized by a high antioxidant potential, components of these protective pathways have been proposed to be used as probes to identify genotypes tolerant or sensitive to abiotic stresses such as

http://dx.doi.org/10.19103/AS.2017.0022.13

cold (Fortunato et al., 2010). Moreover, when environmental variation is perceived as a stress, the response induced at the leaf level can influence metabolism in the fruit, and thus affect its quality (Poiroux-Gonord et al., 2013). It is thus envisaged to identify markers in leaves for plant adaptability to environmental stress, which will constitute indicators of the quality of the future fruits.

Furthermore, certain cultural practices in place such as input supply, phytosanitary treatments or intensive watering are not compatible with a sustainable agriculture and their use needs to be limited. Since ecological and economic benefits are not always compatible, supplementary incomes are necessary to convince farmers to move to sustainable agriculture. Moreover, the culture of coffee plants for bean harvest only provides seasonal employment. To mitigate these problems, value must be added to the coffee crop. At the beginning of the last century, English explorers had observed the traditional use of coffee leaves as infusion in Sumatra and reported its beneficial effects on human health (*New York Times*, 1873). The harvesting and exploitation of the leaves of coffee plants could constitute a supplementary income for farmers, based on their health-promoting benefits since the leaves contain beneficial compounds. Done in a reasonable manner, the leaf harvest will not negatively impact tree growth, bean production and yield.

This chapter provides an inventory of molecules identified to date in the leaves of cultivated coffee trees. A summary of the beneficial effect of the molecules exclusively found in leaves will be given from the perspectives of plant physiology and human health.

2 Characterization of leaf metabolites in cultivated coffee plants

Certainly, because of coffee bean consumption and their great economical interest, all the efforts in biochemical analyses have been essentially carried out on green and roasted beans. Some academic studies have been performed in leaves with the essential goal to decipher the metabolic pathways leading to the synthesis of desirable (or undesirable) compounds in beans (i.e. caffeine, chlorogenic acids and terpenoids). In fact, few data exist concerning the beneficial compounds specifically produced in coffee leaves or in other vegetative tissues differing from fruit and bean.

As an example, in the third volume of a series of books dedicated to the medicinal plants of the world, Ross gives an exhaustive list of the compounds described in the diverse tissues of *Coffea arabica* (Ross, 2005). He documents 360 compounds. However, their presence in leaves concerned 17 compounds that are listed in Table 1. It can be observed that the metabolites described in leaves (or in *in vitro* tissue culture issuing from leaves) are, with the exception of flavonoids and amino acids, derived from the purine alkaloid, terpenoid or hydroxycinnamic acid pathways. These pathways are the most studied in coffee beans, as they lead to the synthesis of caffeine, trigonelline, kahweol, cafestol and chlorogenic acids, compounds that have a great impact on coffee quality. Studies on coffee beans are made difficult by the fact that this material is not available throughout the year and can be obtained only with culture conditions favourable to plant reproduction. Leaves, as a perennial organ with a relatively rapid growth, as well as *in vitro* material, have offered the possibility to realize biochemical studies independent of growth season. They have been used to understand the metabolic pathways existing in beans and more specifically to decipher the metabolism of caffeine (Frischknecht et al.,

Table 1 List of the compounds identified in *Coffea arabica* leaves or tissue culture derived from leaves, their biochemical family and the sources in which they are described (from Ross, 2005)

Compound	Family	Plant material	Sources
Adenine 7 glucosyl	Purines	Suspension culture	Schulthess et al. (1993)
Caffeine	Purine alkaloids	Pericarp, seeds	Higuchi et al. (1995)
Theobromine	Purine alkaloids	Tissue culture	Frischknecht and Baumann (1985)
7-Methylxanthine	Purine alkaloids	Suspension culture	Baumann et al. (1983)
Allantoic acid Allantoin	Ureids: Purine catabolism	Plant	Hoffman et al. (1969)
Kahweol 16-0 Methyl Cafestol	Diterpenoids	Leaves	Kolling-Speer and Speer (1997)
Ursolic acid	Triterpenoids	Leaves	Waller et al. (1980)
Histidine	Amino acids	Plant	Hoffman et al. (1969)
Pipecolic acid (Syn: homoproline) Proline, Hydroxyl	Amino acids	Pericarp, seed	Higuchi et al. (1995)
Hydrolase, nucleotide *N*-Methyl transferase	Proteins	*In vitro*	Waller et al. (1991)
Quercetin glucoside	Flavonoids	Leaves	Gonzalez et al. (1975)
Tannins	Flavonoids	Plant	Atal et al. (1978)
Chlorogenic acids	Hydroxycinnamic acids	Plant	Sondheimer (1958) Shiroya and Hattori (1955)

1986; Nazario and Lovatt, 1993; Ashihara et al., 1996; Mösli Walddhauser et al., 1997) and trigonelline (Zheng et al., 2004) or chlorogenic acids (Colonna, 1986; Mondolot et al., 2006). However, although studies have laid the molecular basis for the biosynthetic pathways of the compounds accumulated in beans, they have not provided the possibility to identify pathways specific to leaves, nor of previously uncharacterized molecules.

With the objective to increase production and coffee quality, agronomical and eco-physiological studies involving leaves have been conducted in the field. In the general belief, wild coffee is an understory tree and subsequently cultivated coffee plants have been selected to be grown in full sunlight. These culture conditions are characterized by higher temperatures, higher light intensities, different light qualities and less water or nitrogen availability than in the natural areas where coffee is grown. These modifications in environmental conditions may have an impact on the plant growth by altering photosynthesis and, influencing the carbon supply and consequently, the plant productivity. Thus, many studies have focused on the effect of light, temperature, drought and nitrogen availability on leaf photosynthesis, photo-oxidative stress and its consequences on plant growth and metabolism (Ramalho et al., 1997; 2000; Barros et al., 1997; Da Matta et al., 1997; Pinheiro et al., 2004; Pompelli et al., 2010).

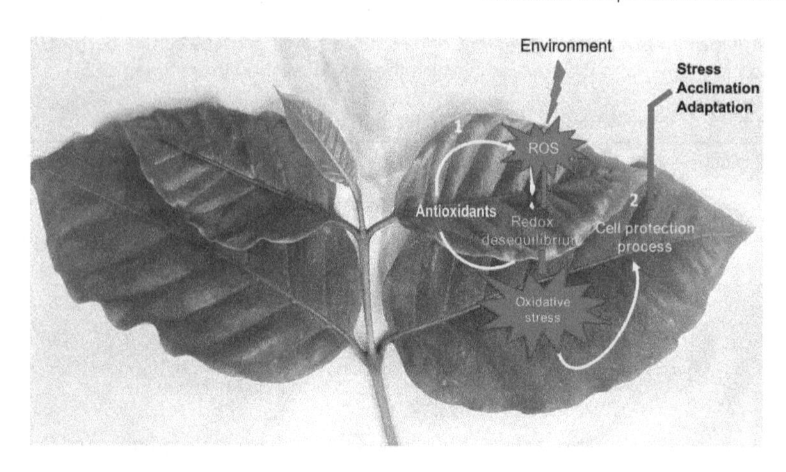

Figure 1 Schematic representation of the events generated in coffee leaf by an environmental stress.

Light stress provokes the accumulation of reactive oxygen species (ROS) that induce an immediate reaction consisting in the production of ROS scavengers in the leaf to mitigate the redox disequilibrium (Step 1, Fig. 1). When the oxidative stress is maintained, a second step is activated (Pathway 2, Fig. 1). It involves the protein kinase signalling pathway (mitogen-activated protein kinase (MAPK) cascade), which regulates the activity of transcription factors (Moustafa et al., 2014; Jalmi and Sinha, 2015). In turn, these proteins regulate the expression of genes involved in metabolic pathways leading to maintenance of cell integrity by limiting damage to lipids, proteins and DNA and the synthesis of compounds intervening in stress response. Each of these factors, irrespective of their level of intervention in the stress response, can be considered as beneficial compounds for the plant by facilitating adaptation to new constraints.

Nevertheless, in coffee leaves, existing studies have dealt essentially with the molecules presenting antioxidant function. Enzymes and antioxidative compounds involved in the first step and phenolics produced during the second step have been described during the response of coffee to cold stress, pathogen attack or light intensity variation (Silva et al., 2002; Campos et al., 2003; Silva et al., 2006; Fortunato et al., 2010; Rodríguez-López et al., 2013). For phenolics, as for the studies concerning the metabolic pathways, only compounds accumulated in beans have been examined.

The discovery of a new molecule in coffee leaves was made possible by a study comparing the leaf biochemistry of wild coffee species for chemotaxonomy classifications (Campa et al., 2010). The biochemical diversity of coffee beans allowed discrimination of wild coffee species according to their geographical origin (Anthony et al., 1993; Campa et al., 2005). The diversity of leaf metabolites was evaluated in more than 20 species of wild coffee plants from East Africa (Mozambicoffea group), West Africa (Eucoffea group) and Madagascar (Mascarocoffea group). Species from the Eucoffea group accumulated more chlorogenic acids than did species from Mascarocoffea group (Fig. 2), consistent with previous studies in beans. As in the case of beans (Rakotomalala, 1992; Campa et al., 2008), uncharacterized alkaloids and methylated compounds are present in leaves of species from the Mascarocoffea group.

Moreover, some species from the Mozambicoffea group were characterized by the presence of a compound having a high yellow fluorescence under UV light (Fig. 3).

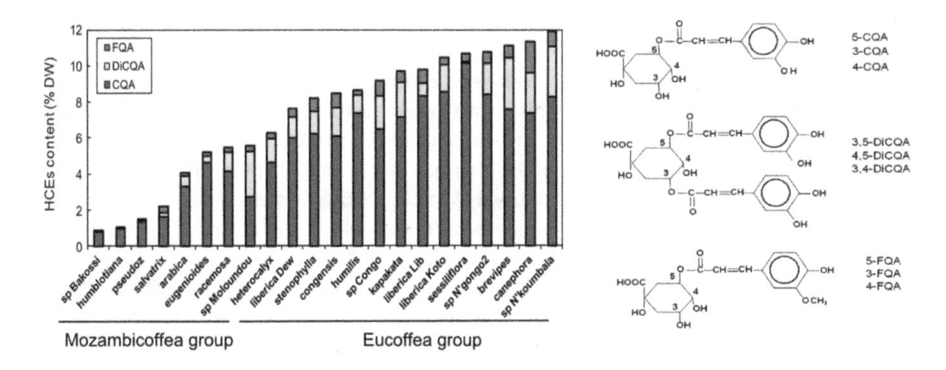

Figure 2 Chlorogenic acid content in green beans of some wild coffee species.

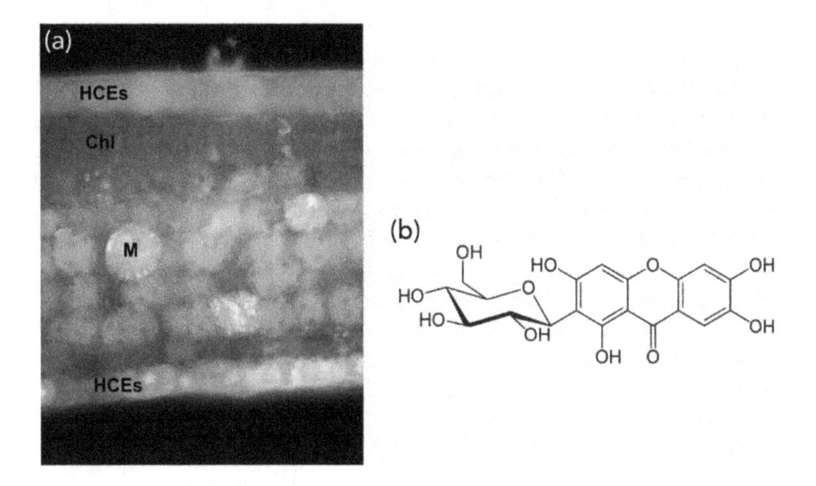

Figure 3 (a) Transversal section of a juvenile leaf of *Coffea arabica*. (b) Chemical structure of the mangiferin, a *C*-glucosyl xanthone.

Its isolation in the wild species, *Coffea pseudozanguebariae*, confirmed that this compound is mangiferin, a *C*-glucosyl xanthone already described in mango and grouped with the benzophenone derivatives (Talamond et al., 2008). The structure, 2-*C*-β-D-glucopyranosyl-1,3,6,7-tetrahydroxyxanthone, was established by Iseda in 1957. Another xanthone is also present and identified as the iso-mangiferin (4-*C*-β-D-glucopyranosyl-1,3,6,7-tetrahydroxanthone), previously isolated from *Anemarrhena asphodeloides* (Aritomi and Kawasaki, 1970).

Xanthones form a large group of more than 500 natural products widely found only in some higher plant families, lichens and fungi. In higher plants, they remain mostly associated with Clusiaceae and Gentianaceae families and occasionally found in phylogenetically distant families such as Iridaceae, Liliaceae, Anacardiaceae, Euphorbiaceae or Verbenaceae. But their presence in Rubiaceae, to which *Coffea* genus belongs, has not been described previously.

The pathway leading to its biosynthesis and that of benzophenone derivatives is as yet unknown in coffee plants, but in *Hypericum* and *Antirrhinum* it occurs via the benzoic acid (BA) biosynthetic route (Abd El-Mawla and Beerhues, 2002; Liu et al., 2003; Long et al., 2009; Moerkercke et al., 2009). This route belongs to the general phenylpropanoid pathway, the common route of most of the phenolic compounds, such as chlorogenic acids, lignins or flavonoids. A branch point is proposed to occur at the cinnamic acid step, the first secondary metabolite of the pathway which is obtained after deamination of the amino acid phenylalanine by phenylalanine ammonia lyase (Fig. 4).

An analysis involving 23 coffee species (represented by 49 genotypes) including nine species from Madagascar evidenced the presence of the mangiferin in eight of the nine species from the Mozambicoffea group (Campa et al., 2012). For species that contain mangiferin, its content in mature leaves varied from 0.026% to 16.3% on a dry weight (DW) basis in *Coffea sessiliflora* and *Coffea salvatrix*, respectively. As observed for other major compounds such as chlorogenic acid or caffeine, the leaf content decreased when leaf aged, as if their synthesis was realized in the apex or in young tissues. The presence of mangiferin is also detected in the fruit pericarp of species accumulating this compound in leaves.

The presence of mangiferin has been recently mentioned in leaves of *C. arabica* cultivated in Brazil (Martins et al., 2014). Its accumulation seemed sensitive to light, the leaves harvested on trees cultivated in full sun having a twofold higher relative content than that of trees cultivated in the shade. Quantification of mangiferin and its isomer, the iso-mangiferin, was recently determined in *C. arabica* leaves from Brazil and Costa Rica. The content varied from 0.067% to 0.497% (DW basis) and mangiferin accounted for 80% or more of the total amount (Trevisan et al., 2016).

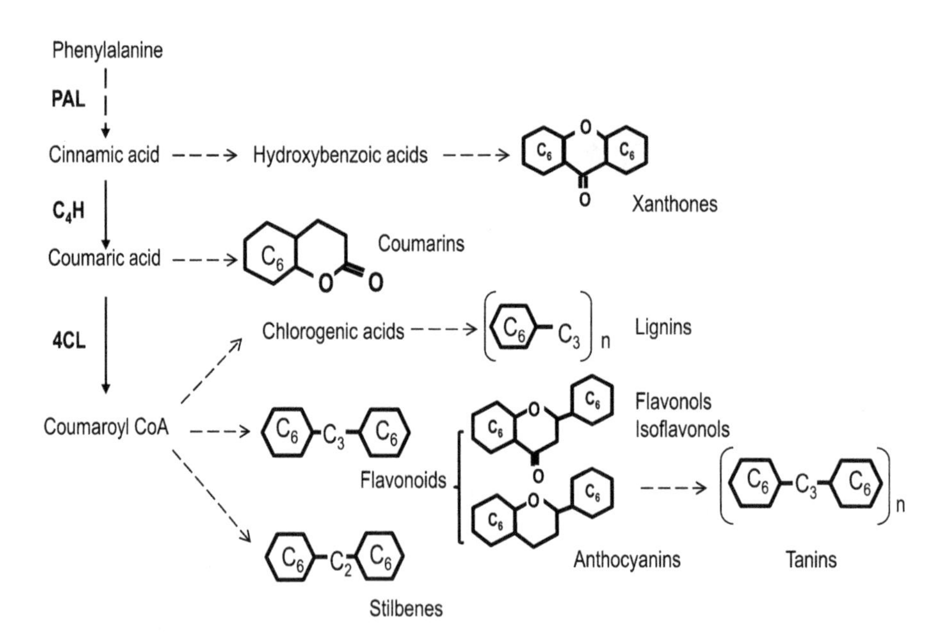

Figure 4 Simplified representation of the general phenylpropanoid pathway.

3 Beneficial compounds for coffee plants

In natural conditions, plants are unable to move and are subject to environmental changes inflicted on them. Factors such as atmospheric changes, climate warming, pathogen attacks or land use are having deep and extensive impacts on their growth and reproduction. To survive these constraints, each species has developed adaptation strategies that allow it to acclimate and ensure its adaptation to new environmental characteristics. Plant secondary metabolites play a central role in these plant adaptations. They intervene (directly or indirectly) in the response, allowing acclimation in case of abiotic modification or resistance in case of pathogen attack. By this way, natural compounds involved in acclimation, adaptation or resistance of plants can be considered as beneficial for plants. As an example, the accumulation of caffeine in very young plant parts of *C. arabica* was thought to be a chemical defence for tissue with a high risk of predation (Frischknecht et al., 1986).

The response of the plant will be adapted to the nature and severity of the stress, and will depend on the ability of the plant (or genotype) to sense, respond and tolerate the stress imposed. However, abiotic or biotic stress results in an oxidative stress at the cellular level, characterized by the intensification of the production of ROS, toxic forms of oxygen that may alter plant metabolism by oxidation of proteins and nucleic acids and by lipid peroxidation. To attenuate this oxidative stress, plants develop a defence system in two principal steps comprising a stress modulation step by antioxidants followed by a general metabolism disorder leading to the synthesis of other antioxidants, generally from the phenolic compounds (see Fig. 1).

Concerning the first step of the stress response, cultivated coffee plants possess antioxidant compounds typically present in all other plants. Lipophilic compounds, such as α-tocopherols and β-carotenes; hydrophilic compounds, such as ascorbate or glutathione; and enzymes (Fig. 5) constitute the antioxidant defence system against ROS. A review of the antioxidant molecules intervening during light stress in mature leaves of Arabica was proposed by Pompelli et al. (2010). These authors noticed that carotenes (α and β) and xanthophylls (neoxanthin, lutein, violaxanthin, antheraxanthin and zeaxanthin) were present but only lutein, antheraxanthin and zeaxanthin seemed to play a role in light response. Chaves et al. (2008) indicated a higher ascorbate pool in sunlit compared to shaded leaves of Arabica trees during rainy and warm season. Both authors studied the activity of the major antioxidant enzymes, that is, superoxide dismutase (SOD; EC 1.15.1.1), ascorbate peroxidase (APX; EC 1.11.1.11), catalase (CAT; 1.11.1.6) and glutathione reductase (GR; EC 1.6.4.2). While present in the leaves, their activity varies slightly depending on the light conditions, suggesting a minor role in the response to light and a low phenotypic plasticity of coffee leaves to varying irradiance.

With regard to the second step of the stress response, no data are available to date concerning the MAPK cascade in coffee leaves. Most of the studies are directed towards the antioxidant compounds involved in this last phase of the stress response and concerned the phenolics. This large family is divided into monophenols; coumarins and hydroxycinnamic acids (HCAs) such as coumaric, caffeic, ferulic or sinapic acids, which can be on simple or complex forms; and polyphenols, flavonoids, stilbenoids and polymers obtained by monomer polymerization that generates lignins, suberins and tannins. Some authors also include the xanthones derived from BA (Fig. 4). In coffee plants, the most accumulated phenolics in green beans and leaves are the chlorogenic acids (Lepelley et al., 2007; Mahesh et al., 2007), antioxidant compounds (Shahidi and Chandrasekara,

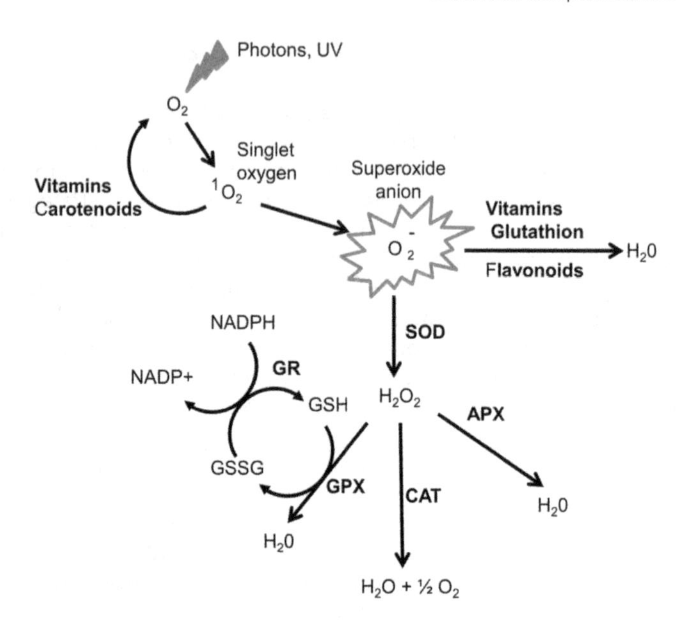

Figure 5 Simplified representation of the antioxidant defence system against one of the ROS: the superoxide anion.

2010) formed by esterification between HCAs and quinic acid. The most prevalent one is the 5-*O*-caffeoylquinic acid (5-CQA), also called chlorogenic acid (Ribereau-Gayon, 1968; Schuster and Herrmann, 1985; Manach et al., 2004).

The low flavonoid content in beans is perhaps the reason why these compounds have been poorly studied in coffee plants. It is only recently that kaempferol, quercetin, catechin and epicatechin have been designed as the major flavonoids in coffee beans and their total content have been estimated as 1.81 mg QE/g (QE: quercetin equivalent) in spent coffee (Mussatto et al., 2011; Panusa et al., 2013; Mussatto, 2015). Some works on pulp and skin of Arabica mature fruits indicated the presence of two anthocyanins, the cyanidine-3-rutinoside and glucoside, in appropriate concentrations to consider coffee as a good source of natural food colorants (Prata and Oliveira 2007; Murthy et al., 2012).

To our knowledge, the first report on flavonoid content in coffee leaves has been published in 2014. By comparing metabolic content of leaves of shaded and full-sun Arabica trees cultivated in south-eastern Brazil, Martins et al. showed that eight flavonoids are accumulated: four glycosylated forms of quercetin and four of kaempferol. The content of quercetin-3-glucosyl-6-rhamnoside (rutin) and kaempferol-3-glucosyl-6′-rhamnoside was particularly increased (x100) in leaves of full-sun plants (Fig. 6). An increased accumulation of the xanthone mangiferin was also observed in high light but at a less extent (x2.3) while chlorogenic acid content evolved differently according to the molecular structure of the compound (isoform or number of esterifications). However, as the content is expressed in relative content, the real concentration in the different flavonoid compounds is still unknown.

The presence of flavonoids in coffee leaves is not surprising, as the synthesis of flavonoids, molecules having a defensive role against excess heat and UV light, is a

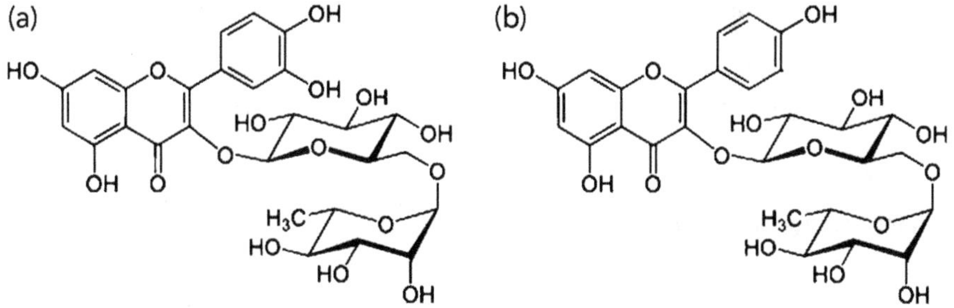

Figure 6 Two flavonoids that are principally accumulated in full-sun leaves of *Coffea arabica*. (a) Rutin, quercetin-3-glucosyl-6-rhamnoside; (b) kaempferol-3-glucosyl-6-rhamnoside.

characteristic of the vascular plants. The emergence of this biosynthesis pathway allowed plants to colonize Earth's terrestrial environment nearly 500 million years ago (Mouradov and Spangenberg, 2014). Flavonoids have additional beneficial roles for plant survival. As a consequence of their high antioxidant power, flavonoids have been shown to be involved in most of the stages of the plant life cycle, ensuring plant resistance (Treutter, 2006) and protection against insect predation and microbes, plant reproduction by attracting insects for pollination (Winkel-Shirley, 2001), favouring pollen germination (Ferreyra et al., 2012) and biological dialogs with the rhizosphere (Weston and Mathesius, 2013), and participating in the developmental regulation by their involvement in auxin transport and catabolism (Brown et al., 2001).

Considering the coffee leaf, there is a surprising lack of data on flavonoid content, with the exception of the reports of Domingues Junior et al. (2012) on the anthocyanins in young leaves and Martins et al. (2014). This may be due to the low capacity of coffee leaves to accumulate flavonoids, which would make them difficult to detect. This hypothesis is consistent with the weak ability of coffee tree to protect against UV and heat stress and would explain the preference of wild coffee trees for growing in the undergrowth of tropical forests. Another interesting point concerning coffee leaves is their high content in chlorogenic acids, phenolics of the lignin pathway (Fig. 4) conferring beneficial roles in plant cells. Soluble esters formed between hydroxycinnamic acids and quinic acids, these compounds intervene in plant protection against abiotic (Grace et al., 1998; Clé et al., 2008) and biotic stresses (Niggeweg et al., 2004; Leiss et al., 2009). Largely accumulated in coffee beans, particularly in that of Robusta, chlorogenic acids also constitute the most important phenolic family present in coffee leaves. However, they are much less abundant than in green beans (Campa et al., 2012) and in contrast to what is observed in the bean, the leaf content of *C. arabica* (var. Laurina) is higher than of *Coffea canephora* (3.9% and 2.4% DW leaf, respectively). Their content decreases with leaf age, particularly for the 3,5-dicaffeoylquinic acid (3,5-DiCQA), a more stronger antioxidant than 5-caffeoylquinic acid (5-CQA) (Mahesh et al., 2007).

Another interesting antioxidant compound is mangiferin. Absent in Robusta leaves, this compound, initially isolated from mango (Barretto et al., 2008), provides antioxidant and antimicrobial protection upon biotic stress in *Hypericum* (Franklin et al., 2009). In a Madagascan shrub, *Aphloia theiformis*, abundant leaf accumulation is observed in plants subject to high light conditions (Danthu et al., 2010). In coffee plants, it can play a role in plant adaptation, as its presence was documented in species able to colonize high altitudes (Campa et al., 2012). However, for *Swertia,* altitude is not considered as the greatest parameter that influences mangiferin accumulation in leaves (Yang et al., 2004; Yang et al., 2005). Mangiferin is perhaps substituting the role of the poorly represented flavonoids, giving tissue protection against UV or intense radiation.

4 Beneficial compounds for humans

Despite the development of extensive research on the effects of coffee drink consumption on human health, the debate continues as to the beneficial effects of the presence of certain metabolites in the drink. Caffeine is certainly the compound whose effects are most debated, with those of trigonelline and the diterpenes, kahweol and cafestol. Issues concerning the presence of chlorogenic acids in coffee are more focused on the contribution to coffee quality rather than the impact on human health. The compounds present in the leaves of cultivated coffee species are nearly identical to those found in fruits (except the presence of mangiferin in Arabica). Therefore, similar questions can be asked about the beneficial effects of the molecules present in coffee leaves as well as beans, if we envisage to use the leaves for beverage.

Coffee leaves, quasi-exclusively *C. arabica*, are traditionally used in some countries, and different uses in folk medicine have been summarized by Ross (2005). The leaves or their decoction are used against headache in Haiti and Nicaragua. Decoction of grilled leaves are taken orally for anaemia, oedema, asthenia and rage in Haiti. Made into a poultice, leaves are used treat fever in Mexico and hot water extracts are taken orally for stomach pain in Nicaragua, and against cough in flu and lung ailments in Peru. Surprisingly, no data exist on a potential use in folk medicine in Africa, the geographical zone of origin of Arabica coffee.

Coffee leaves are also used as infusion in Indonesia, as maté or tea, and the drink has been called 'coffee leaf tea' in reports (*New York Times*, 1873; *The British Medical Journal*, 1876). In the Global Coffee Report (2012), Davis reported that coffee leaf tea was consumed in Jamaica, India, Java and Sumatra and also in Ethiopia and other parts of Africa and that a patent was obtained in the 1850s by Dr Gardner of Ceylon for its preparation. Consumption ceased with the end of the culture of Arabica, when the plantations were devastated by leaf rust. Surprisingly, this use has not returned with the culture of Robusta. Except for the differences in content for caffeine and chlorogenic acids, the leaves of the species differ essentially by the presence of mangiferin. Is this compound that makes the coffee leaf tea obtained from Arabica drinkable and with beneficial characteristics compared to Robusta?

Mangiferin is also named euxanthogen, alpizarin, aphloïol, chimonin, hedysarid or shamimin, according to the author who isolated the molecule and the plant from which it was obtained. This compound has been largely studied from the beginning of the last century, as it seems to have numerous pharmacological properties. It is described as a strong antioxidant that displays anti-allergic (Rivera et al., 2006), anti-bacterial, anti-viral (Yoosook et al., 2000), anti-inflammatory, anti-nociceptive (Garrido et al., 2004), anti-diabetic (Miura et al., 2001), anti-allergic (Garcia et al., 2003), anti-hyperlipidaemic (Muruganandan et al., 2005), anti-carcinogenic (Pinto et al., 2005) and neuroprotective activities (Campos-Esparza et al., 2009). In 2007, two patents have been filed, one proposing xanthones for treating diabetes (Wada, 2007) and the other indicating mangiferin as an agent treating and preventing neurodegenerative diseases and ageing symptoms (Matute et al., 2007). An updated account of the patents published on important therapeutic properties of mangiferin is given by Khurana et al. (2016).

Research is aimed at determining the mode of action of the mangiferin and the derivation of therapeutic molecules (King et al., 1999; Han et al., 2015). Concerning the neuroprotective activity of mangiferin, *in vitro* research using neuronal cultures from rats attributed the effect of the compound to its ability to reduce the formation of ROS, to activate the enzymatic antioxidant system (SOD, CAT enzymes) and to restore the mitochondrial membrane potential (Campos-Esparza et al., 2009). Dimitrov et al. (2011) showed an *in vivo* antidepressant effect of mangiferin on mice and attributed this effect to the inhibition of the monoamine oxidases A (MAO_A), oxidoreductases involved in the degradation of the monoamines, neurotransmitters and neuroregulators intervening in the regulation of most of the cognitive processes. A protective effect on cerebral ischaemia-reperfusion injury was shown by Yang et al. (2016). Among different effects, mangiferin upregulated the activities of SOD, glutathione (GSH) and that of an inducible transcription factor regulating multiple cellular antioxidant systems that limit oxidative stress, the erythroid 2-related factor (Nrf2).

Studying the molecular mechanisms involved in the anticancer effect, Shoji et al. (2011) and Takeda et al. (2016) showed that mangiferin induced apoptosis *in vitro*, in acute myeloid leukaemia and multiple myeloma cell lines, respectively, by inhibiting the nuclear translocation of the nuclear factor kappa B (NF-κB). This transcription factor is a protein complex that controls a variety of cellular processes, such as DNA transcription, cytokine production or cell survival. Involved in cellular responses to stimuli such as stress, NF-κB is found in almost all animal cell types. When inactive, it is sequestered in a complex form in the cytoplasm. By the increase of activity of a NF-κB-inducing kinase (NIK), a serine/threonine protein kinase from the MAPKs, NF-κB is translocated to the nucleus where it plays a central role in pathogenicity. The different mechanisms of action of mangiferin on cancer development were reviewed by Gold-Smith et al. (2016).

Other *in vivo* and *in vitro* studies have shown a similar mode of action in case of asthma, arthritis, bronchitis and stress (Rivera et al., 2011; Tsubaki et al., 2015; Kumar et al., 2003; Márquez et al., 2012). As NF-κB is involved in the activation of inflammatory mediators, the anti-inflammatory effect of mangiferin was also assessed in colitis (Jeong et al., 2014). *In vitro* assays of different concentrations of mangiferin on peritoneal macrophages of treated mice indicated that the compound limits inflammatory disease by regulating NF-κB and MAPK signalling pathways through the inhibition of an interleukin-1 receptor-associated kinase (IRAK1) and/or MAPKs phosphorylation. Phosphorylated IRAK1 or MAPKs in macrophages activates a multimeric protein complex, which in turn activates NF-κB and the expression of pro-inflammatory cytokines.

This molecular mechanism is also proposed studying the protecting role of mangiferin in diabetic nephropathy and myocardial insults (Pal et al., 2014; Suchal et al., 2016). In rat, Pal et al. (2014) showed that mangiferin protects cells from apoptotic death by inhibiting all the changes induced by the oxidative stress created by diabetic nephropathy. In particular, the molecular structure of mangiferin contains four phenolic H atoms (Fig. 3) which allows donation of two atoms to free radicals (e.g. ROS) in order to form stable phenoxyl radicals. As confirmed by Mendoza-Sarmiento et al. (2016), this characteristic makes mangiferin an excellent antioxidant.

5 Case study: Wize Monkey

In early 2013, international media coverage (Gray, 2013) highlighted the beneficial compounds from coffee leaves based on the study 'A survey of mangiferin and hydroxycinnamic acid ester accumulation in coffee (*Coffea*) leaves: biological implications and uses' (Campa et al., 2012). This article inspired two business school students, Arnaud Petitvallet and Max Rivest, to examine the existing use of coffee leaves and introduce this product (Fig. 7) to the world market.

They discovered an historical use of the coffee leaf in the Harari region in Ethiopia, where the infusion from coffee leaves has been traditionally consumed, and in Sumatra, Indonesia (Simmonds, 1864). They discovered that coffee communities traditionally used sun-dried Arabica leaves, to be then infused and consumed with milk and spices. The consumption of the coffee leaf was, however, not recorded in Latin America.

The coffee plant does not produce any coffee beans before it reaches three years of age, and can only be harvested for a period ranging between 3 and 4 months each year, depending on the geographic and climatic conditions, with some regions (Colombia) being able to harvest twice a year. Moreover, diseases, pests and climatic conditions are additional factors that can impact the quality and the yield of the coffee bean.

Farmers do not have a degree of control over the resale price for coffee, creating unpredictable revenue for the year. The short window of bean harvest provides a large influx of money that is often depleted before the end of the year and leaving the farmers with a 3-month period of hunger, referred as 'the thin months' (Bacon, 2014).

With the majority of farms being small-scale farms (under 2 acres), farmers have to organize in cooperatives to reinforce their bargaining power with buyers. The coffee price index makes it hard for farmers to get paid at the right price, and can create disastrous consequences when reaching a low point.

One of the worst example of this price volatility was the 'tuna crisis' that happened in the early 2000s in Nicaragua. Unable to reach a sustainable price point, farmers in the region were not able to harvest coffee beans, creating a major social crisis in the coffee lands, with children dying, farmers leaving coffee farms and marching towards the capital city to ask for food and medicine (Nicaragua, 2003). Countries like Nicaragua also face an important social migration effect at the end of the harvest, with an average 80% of the seasonal workers being sent home as soon as the harvest ends. The country's industrial infrastructure is poorly developed, forcing the workers and their families to migrate towards neighbouring countries or travel to other regions.

After learning about this situation and gathering more information on the coffee leaf, Petitvallet and Rivest decided to focus their Business Plan Competition project on creating a line of tea products based on the coffee leaf under the name 'Wize Monkey'. In June 2013, they finished 2nd in their school's competition and went to Nicaragua with the goal of sourcing coffee leaves from farmers and locally processing them.

Nicaraguan coffee farmers were not using any of the leaf on the coffee farms. The concept of infusing dried coffee leaves was unknown in the region, despite a popular knowledge of plant-based remedy and medicine. Most of the farmers welcomed the idea of additional revenue, especially considering the low prices of coffee beans at the time, but expressed their doubt in sustaining a proper coffee bean production with a leaf harvest. The majority of coffee farms producing for cooperatives were poorly managed due to lack of proper agronomic training and lack of resources for maintenance. Most coffee

Figure 7 One packaging of the coffee leaf tea from Wize Monkey.

plants were lacking nutrients and were underdeveloped, making coffee leaf harvest very challenging.

Wize Monkey then introduced the concept to private estates in Nicaragua, with larger centralized operations. Coffee plants located on those estates tend to receive better treatment, nutrients and typically produce more beans and leaves. Working with local farmers, Wize Monkey identified several methods to harvest coffee leaves without damaging the plant. To harvest in a sustainable manner, it was decided to collect mature coffee plants, starting roughly 2 months after the end of the coffee harvest to give time for the plant to grow new shoots.

The leaf harvest, combined with the processing, handling and shipping, is now able to ensure a consistent activity on the coffee farm during this off-season, with encouraging early results. Wize Monkey was able to create 30 jobs for local workers during the leaf harvest period in late summer 2015, and over 100 during its 2016 harvest. With the perspective of a stable revenue and employment on the coffee farm, the families are now able to send their kids to school full time. Wize Monkey partners with local NGOs building schools on coffee farms to educate more children and push higher the age of education completion.

As the domestic and international demand is growing for coffee leaf-based product due to its promising health benefits and its taste, Wize Monkey hopes to bring 1000 stable jobs in the coffee communities of Nicaragua by 2020, while developing local knowledge on the leaf and give the keys to sustainable agriculture and year-round revenue.

6 Conclusion

Further research on beneficial compounds in leaves from coffee plants can contribute to offset the decline of coffee lands. It is estimated that by 2050, 50% of the Arabica coffee lands will not produce beans anymore, due to climate change effect (Bunn et al., 2015).

Selecting plants able to accumulate compounds with antioxidant potential may favour a better tolerance of Arabica coffee plants to changing climatic constraints and, moreover, a better organoleptic quality of the beans. Thus, as it has been shown for *Citrus sinensis*, the ability of the tree leaves to synthesize carotenoids in response to photo-oxidative stress impacts the metabolite content of the orange fruit and, by this way, its quality (Poiroux-Gonord et al., 2013). Evaluating the content in antioxidant metabolites in coffee leaves may then constitute a new tool to contribute to the selection of plants able to give sustainable crop production. Research focused on elucidating the biosynthetic pathways of these compounds and their regulation will help in understanding the complex cellular signalling networks set by the leaves in response to external stimuli and finding the culture conditions favourable to a sustainable crop production.

Moreover, if the coffee leaf is perceived to be beneficial for human health, the coffee leaf might represent long-term solution for thousands of producers in all coffee-producing countries threatened by the effects of climate change. In this case, harvesting leaves will constitute an alternative source of income from the same trees, but available all along the year. Mangiferin appears as one example of these promising compounds. Accumulated by the leaves of one of the two major cultivated coffee species, it is present in infusions from dried leaves that are consumed by humans. When isolated from Arabica leaves, it seems that it will become a valuable pharmacological substance, demonstrating activity against diverse human diseases.

7 Future trends

One of the first objectives in research will be to understand the fine molecular mechanisms involved in the response of coffee to biotic and abiotic stresses. The answer to this question should provide new tools for selecting coffee plants that are tolerant to the multiple stresses associated with climate change without altering their production potential and quality.

In this objective, it will be necessary to develop genetic and genomic approaches to identify the pathways involved in stress response in coffee plants, in particular those constituting and governing the MAPK cascades. As shown by Lee et al. (2016), MAPK cascades are present in guard cells and play important roles in the signalling networks in response to biotic and abiotic stresses. Identification of the active MAPKs and their roles could be done by comparing their expression level and that of the transcription factors they control among different situations of stress. Candidate genes can be easily obtained by consulting the published whole-genome sequence from *C. canephora* (Denoeud et al., 2014).

Moreover, among the different potential antioxidant molecules accumulated in leaves, the signalling molecules and active antioxidants have to be identified. Different studies on the effects of mangiferin on different pathologies affecting human health have indicated its intervening at different levels of the stress response. It may act directly as an antioxidant but also by stimulating or decreasing the activity of some proteins involved in the cell response, such as MAPKs or transcription factors. In plants, the precise role and the modes of action of the mangiferin have not been studied yet. Then, it will be particularly interesting to search what effects mangiferin exerts in plants and if this compound plays the same role as in animal cells in steps 1 and 2 of the stress response, as summarized in the Fig. 8. In particular, it will be interesting to verify the inhibiting (or stimulating) role of mangiferin on transcription factors governed in plants by MAPK cascades.

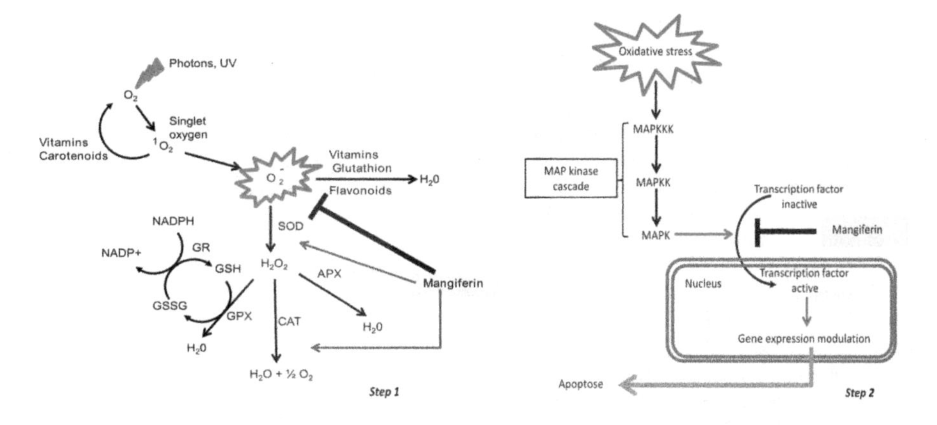

Figure 8 Hypothetical mode of action for mangiferin in plant cells, based on the results obtained on animal cells.

Another trend will be to take advantage of the great diversity observed in wild coffee species in order to search species accumulating compounds demonstrating beneficial interest for plants and/or humans. More especially, the biochemical diversity of the leaves (and fruit when unknown) of the species from the Mascarocoffea group and from the subgenus *Psilanthus*, a group of coffee plant from Asia that have been recently added to the *Coffea* genus, has to be precisely investigated. Using metabolomics, it will be possible to identify novel molecules, which can be tested for their functions in plants and human health. Metabolites allowing development of coffee culture in more sustainable conditions (less watering, less input addition and less pesticide) will be targeted.

8 Where to look for further information

For some details about mangiferin, it will be useful to read the recent paper of Khurana et al. (2016) that proposes a summary of mangiferin effects and patents.

For a better knowledge on MAPK cascades and the channels they govern, the papers of Moustafa in Cell Press (2014), Jalmi and Sinha (2015) and Lee et al. (2016) are particularly interesting.

To have some details about metabolomics, consult the site of the Réseau Francophone de Métabolomique et Fluxomique: http://www.rfmf.fr/

9 References

Abd El-Mawla, A. M. A. and Beerhues, L. (2002). 'Benzoic acid biosynthesis in cell cultures of *Hypericum androsaemum*', *Planta*, 214, 727–33.

Adugna, G., Yadessa, G. B. and Ocho, F. L. (2016). 'Threats of climate change on arabica coffee (*Coffea arabica* L.) in its center of origin Ethiopia'. In Plant and Animal Genome Conference XXIV, San Diego, CA.

Anon. (1873). 'Coffee-leaf Tea', *NYTimes.com*, 14 September.
Anon. (1876). 'Coffee-leaf tea', *The British Medical Journal*, 25 November, p. 691.
Anon. (2012). *'The power of the leaf', Global Coffee Report*, November 2012, pp. 44–6.
Anthony, F., Clifford, M. N. and Noirot, M. (1993). 'Biochemical diversity in the genus *Coffea* L.: chlorogenic acids, caffeine and mozambioside contents', *Genetic Resources and Crop Evolution*, 40, 61–70.
Aritomi, M. and Kawasaki, T. (1970). 'A new xanthone C-glucoside, position isomer of mangiferin, from *Anemarrhena asphodeloides* Bunge', *Chemical and Pharmaceutical Bulletin*, 18, 2327–33.
Ashihara, H., Monteiro, A. M., Gillies, F. M. and Crozier, A. (1996). 'Biosynthesis of caffeine in leaves of coffee', *Plant Physiology*, 111, 747–53.
Atal, C. K., Srivasava, J. B., Wali, B. K., Chakravarty, R. B., Bhawan, B. N. and Rastogi R. P. (1978). 'Screening of Indian plants for biological activity. Part VIII', *Indian Journal of Experimental Biology*, 16, 330–49.
Bacon, C., Sundstrom, W. A., Flores Gomez, M. A., Mendez, V. E., Santos, R., Goldoftas, B. and Dougherty, I. (2014). 'Explaining the "hungry farmer paradox": Smallholders and fair trade cooperatives navigate seasonality and change in Nicaragua's corn and coffee markets', *Global Environmental Change*, 25, 133–49.
Barreto, J. C., Trevisan, M. T. S., Hull, W. E., Erben, G., de Brito, E. S., Pfundstein, B., Würtele, G., Spiegelhalder, B. and Owen, R. W. (2008). 'Characterization and quantitation of polyphenolic compounds in bark, kernel, leaves, and peel of mango (*Mangifera indica* L.)', *Journal of Agricultural and Food Chemistry*, 56, 5599–610.
Barros, R. S., da S.e Mota, J. W., da Matta F. M. and Maestri, M. (1997). 'Decline of vegetative growth in *Coffea arabica* L. in relation to leaf temperature, water potential and stomatal conductance', *Field Crops Research*, 54, 65–72.
Baumann, T. W., Koetz, T. W. and Morah, R. (1983). 'N-methyltransferase activities in suspension cultures of *Coffea arabica*', *Plant Cell Reports*, 2 (1), 33–5.
Brown, D. E., Rashotte, A. M., Murphy, A. S., Normanly, J., Tague, B. W., Peer, W. A., Taiz, L. and Muday, G. K. (2001). 'Flavonoids act as negative regulators of auxin transport *in vivo* in Arabidopsis', *Plant Physiology*, 126 (2), 524–35. doi: http://dx.doi.org/10.1104/pp.126.2.524.
Bunn, C., Läderach, P., Ovalle Rivera, O. and Kirschke, D. (2015). 'A bitter cup: climate change profile of global production of Arabica and Robusta coffee', *Climatic Change*, 129 (1), 89–101.
Bustos, A. G. P. (2007). 'Allelochemical effects of aromatic species intercropped with coffee (*Coffea arabica* L.) in Puebla, Mexico', PhD thesis, Bogota, Columbia, pp. 100.
Campa, C., Doulbeau, S., Dussert, S., Hamon, S. and Noirot, M. (2005). 'Qualitative relationship between caffeine and chlorogenic acid contents among wild *Coffea* species', *Food Chemistry*, 93, 135–9.
Campa, C., Rakotomalala, J. J., de Kochko, A. and Hamon, S. (2008). 'Chlorogenic acid diversity in green beans of wild coffee species', *Advances in Plant Physiology*, 10, 421–37.
Campa, C., Mathonnet, N. and Rakotomalala, J. J. (2010). 'Composés phénoliques et biodiversité dans le genre *Coffea*', In *Ecologie 2010*, colloque National d'Ecologie scientifique, 2–4 Septembre 2010, Montpellier, p. 216.
Campa, C., Mondolot, L., Rakotondravao, A., Bidel, L. P. R., Gargadennec, A., Couturon, E., La Fisca, P., Rakotomalala, J-J., Jay-Allemand, C. and Davis, A. P. (2012). 'A survey of mangiferin and hydroxycinnamic acid ester accumulation in coffee (*Coffea*) leaves: biological implications and uses', *Annals of Botany*, 110, 595–613.
Campos, P. S., Quartin, V., Ramalho, J. C. and Nunes M. A. (2003). 'Electrolyte leakage and lipid degradation account for cold sensitivity in leaves of *Coffea* sp. plants', *Journal of Plant Physiology* 160, 283–92.
Campos-Esparza, M. R., Sánchez-Gómez, M. V. and Matute, C. (2009). 'Molecular mechanisms of neuroprotection by two natural antioxidant polyphenols', *Cell Calcium*, 45 (4), 358–68. doi:10.1016/j.ceca.2008.12.007.
Chaves, A. R. M., Ten-Caten, A., Pinheiro, H. A., Ribeiro, A. and DaMatta, F. M. (2008). 'Seasonal changes in photoprotective mechanisms of leaves from shaded and unshaded field-grown coffee (*Coffea arabica* L.) trees', *Trees*, 22, 351–61.

Clé, C., Hill, L. M., Niggeweg, R., Martin, C. R., Guisez, Y., Prinsen, E. and Jansen, M. A. (2008). 'Modulation of chlorogenic acid biosynthesis in *Solanum lycopersicum*; consequences for phenolic accumulation and UV-tolerance', *Phytochemistry*, 69 (11), 2149–56.

Colonna, J. P. (1986). 'Biosynthèse et renouvellement de l'acide chlorogénique et des depsides voisins dans le genre *Coffea*. II. Incorporation de la radioactivité de la L-phénylalanine-14C dans l'acide chlorogénique des feuilles de caféier, en présence ou non de compétiteurs isotopiques', *Café Cacao Thé*, XXX (4), 247–58.

Da Matta, F. M., Maestri, M., Mosquim, P. R. and Barros, R. S. (1997). 'Photosynthesis in coffee (*Coffea arabica* and *C. canephora*) as affected by winter and summer conditions', *Plant Science*, 128, 43–50.

Danthu, P., Lubrano, C., Flavet, L., Rahajanirina, V., Behra, O., Fromageot, C., Rabevohitra, R. and Roger, E. (2010). 'Biological factors influencing production of xanthones in *Aphloia theiformis*', *Chemistry and Biodiversity*, 7, 140–50.

Denoeud, F., Carretero-Paulet, L., Dereeper, A., Droc, G., Guyot, R., Pietrella, M., Zheng, C., Alberti, A., Anthony, F., Aprea, G., Aury, J.-M., Bento, P., Bernard, M., Bocs, S., Campa, C., Cenci, A., Combes, M.-C., Crouzillat, D., Da Silva, C., Daddiego, L., De Bellis, F., Dussert, S., Garsmeur, O., Gayraud, T., Guignon, V., Jahn, K., Jamilloux, V., Joët, T., Labadie, K., Lan, T., Leclercq, J., Lepelley, M., Leroy, T., Li, L.-T., Librado, P., Lopez, L., Muñoz, A., Noel, B., Pallavicini, A., Perrotta, G., Poncet, V., Pot, D., Priyono, R. M., Rouard, M., Rozas, J., Tranchant-Dubreuil, C., VanBuren, R., Zhang, Q., Andrade, A. C., Argout, X., Bertrand, B., de Kochko, A., Graziosi, G., Henry, R. J., Jayarama, M. R., Nagai, C., Rounsley, S., Sankoff, D., Giuliano, G., Albert, V. A., Wincker, P. and Lashermes, P. (2014). 'The coffee genome provides insight into the convergent evolution of caffeine biosynthesis', *Science*, 345, 1181–4. doi:10.1126/science1255274.

Dimitrov, M., Nikolova, I., Benbasat, N., Kitanov, G. and Danchev, N. (2011). 'Acute Toxicity, Antidepressive and Mao Inhibitory Activity of Mangiferin Isolated from *Hypericum Aucheri*', *Biotechnology & Biotechnological Equipment*, 25 (4), 2668–71.

Domingues Junior, A. P., Shimizu, M. M., Moura, J. C., Catharino, R. R., Ramos, R. A., Ribeiro, R. V. and Mazzafera, P. (2012). 'Looking for the physiological role of anthocyanins in the leaves of *Coffea arabica*', *Photochemistry and Photobiology*, 88, 928–37.

Elad, Y. and Pertot, I. (2014). 'Climate change impacts on plant pathogens and plant diseases', *Journal of Crop Improvement*, 28 (1), 99–139.

Ferreyra, M. L. F., Rius, S. P. and Casati, P. (2012). 'Flavonoids: biosynthesis, biological functions, and biotechnological applications', *Frontiers in Plant Science*, 3, 222. doi:10.3389/fpls.2012.00222.

Fortunato, A. S., Lidon, F. C., Batista-Santos, P., Leitão, A. E., Pais, I. P., Ribeiro, A. I. and Ramalho, J. C. (2010). 'Biochemical and molecular characterization of the antioxidative system of *Coffea* sp. under cold conditions in genotypes with contrasting tolerance', *Journal of Plant Physiology*, 167 (5), 333–42.

Franklin, G., Conceição, L. F. R., Kombrink, E. and Dias, A. C. P. (2009). 'Xanthone biosynthesis in *Hypericum perforatum* cells provides antioxidant and antimicrobial protection upon biotic stress'. *Phytochemistry*, 70, 60–8.

Frischknecht, P. M. and Baumann, T. W. (1985). 'Stress induced formation of purine alkaloids in plant tissue culture of *Coffea arabica*', *Phytochemistry*, 24 (10), 2255–7.

Frischknecht, P. M., Ulmer-Dufek, J. and Baumann, T. W. (1986). 'Purine alkaloid formation in buds and developing leaflets of *Coffea arabica*: expression of an optimal defence strategy?' *Phytochemistry*, 25 (3), 613–16.

Garcia, D., Escalante, M., Delgado, R., Ubeira, F. M. and Leiro, J. (2003). 'Anthelminthic and antiallergic activities of *Mangifera indica* L. stem bark components vimang and mangiferin', *Phythotherapy Research*, 17, 1203–8.

Garrido, G., González, D., Lemus, Y., García, D., Lodeiro, L., Quintero, G., Delporte, C., Núñez-Sellés, A. J. and Delgado, R. (2004). '*In vivo* and *in vitro* anti-inflammatory activity of *Mangifera indica* L. extract (VIMANG)', *Pharmacological Research*, 50 (2), 143–9.

Gold-Smith, F., Fernandez, A. and Bishop, K. (2016). 'Mangiferin and cancer: mechanisms of action', *Nutrients*, 8 (7). pii: E396. doi:10.3390/nu8070396. Review.

Gonzalez, J., Noriega, R. and Sandoval, R. (1975). 'Contribution to the study of flavonoids of coffee tree (*Coffea*) leaves', *Revista Colombiana de Química*, 5, 85.

Grace, S. C., Logan, B. A. and Adams III W. W. (1998). 'Seasonal differences in foliar content of chlorogenic acid, a phenylpropanoid antioxidant, in *Mahonia repens*', *Plant Cell Environmnet*, 21, 513–21.

Gray, R. (2013). 'Tea made from coffee leaves found to beneficial for health', *The Telegraph, telegraph.co.uk*. 13 January 2013.

Green, P. W. C., Davis, A. P., Cossé, A. A. and Vega, F. E. (2015). 'Can coffee chemical compounds and insecticidal plants be harnessed for control of major coffee pests?' *Journal of Agricultural and Food Chemistry*, 63, 9427–34.

Han, J., Yi, J., Liang, F., Jiang, B., Xiao, Y., Gao, S., Yang, N., Hu, H., Xie, W. F and Chen, W. (2015). 'X-3, a mangiferin derivative, stimulates AMP-activated protein kinase and reduces hyperglycemia and obesity in db/db mice', *Molecular and Cellular Endocrinology*, 405, 63–73. doi:10.1016/j.mce.2015.02.008.

Higuchi, K., Suzuki, T. and Ashihara, H. (1995). 'Pipecolic acid from the developing fruits (pericarp and seeds) of *Coffea arabica* and *Camellia sinensis*'. In: Colloque Scientifique International sur le Café (Comptes Rendus), Vol. 1: Seizième Colloque International sur le Café, 16th edn, 16, 389–95.

Hoffman, E., Schlee, D. and Reinbothe, H. (1969). 'On the occurrence and distribution of allantoin in Boraginaceae', *Flora of a Physiology Biochemistry* (Jena), 159, 510–18.

Iseda, S. (1957). 'Mangiferin, the coloring matter of *Mangifera indica*. IV- Isolation of 1,3,6,7-tetrahydroxanthone and the skeletal structure of mangiferin', *Bulletin of the Chemical Society of Japan*, 30, 625–9.

Jalmi, S. K. and Sinha, A. K. (2015). 'ROS mediated MAPK signaling in abiotic and biotic stress- striking similarities and differences', *Frontiers in Plant Science*, 6, 769. doi:10.3389/fpls.2015.00769.

Jeong, J. J., Jang, S. E., Hyam, S. R., Han, M. J. and Kim, D. H. (2014). 'Mangiferin ameliorates colitis by inhibiting IRAK1 phosphorylation in NF-κB and MAPK pathways', *European Journal of Pharmacology*, 740, 652–61. doi:10.1016/j.ejphar.2014.06.013.

Khurana, R. K., Kaur, R., Lohan, S., Singh, K. K. and Singh, B. (2016), 'Mangiferin: a promising anticancer bioactive', *Pharmaceutical Patent Analyst*, 5 (3), 169–81. doi:10.4155/ppa-2016–0003.

King, P. J., Ma, G., Miao, W., Jia, Q., McDougall, B. R., Reinecke, M. G., Cornell, C., Kuan, J., Kim, T. R. and Robinson, W. E. J. (1999). 'Structure activity relationships: analogues of the dicaffeoylquinic and dicaffeoyltartaric acids as potent inhibitors of human immunodeficiency virus type 1 integrase and replication', *Journal of Medicinal Chemistry*, 42 (3), 497–509.

Kolling-Speer, I. and Speer, K. (1997). 'Diterpenes in coffee leaves', Colloq Sci Int Café, 17 (15), 1–154.

Kumar, I. V. M., Paul, B. N., Asthana, R., Saxena, A., Mehrotra, S. and Rajan, G. (2003). 'Swertia chirayita Mediated Modulation of Interleukin-1β Interleukin-6, Interleukin-10, Interferon-γ, and Tumor Necrosis Factor-α in Arthritic Mice', *Immunopharmacology and Immunotoxicology*, 25 (4), 573–83. doi:10.1081/IPH-120026442.

Lee, Y., Kim, Y. J., Kim, M-H. and Kwak, J. M. (2016). 'MAPK cascades in guard cell signal transduction', *Frontiers in Plant Science*, 7, 80. doi:10.3389/fpls.2016.00080.

Leiss, K. A., Maltese, F., Choi, Y. H., Verpoorte, R. and Klinkhamer, P. G. (2009). 'Identification of chlorogenic acid as a resistance factor for thrips in chrysanthemum', *Plant Physiology*, 150 (3), 1567–75.

Lepelley, M., Cheminade, G., Tremillon, N., Simkin, A., Caillet, V, and McCarthy, J. (2007). 'Chlorogenic acid synthesis in coffee: an analysis of CGA content and real-time RT-PCR expression of HCT, HQT, C3H1, and CCoAOMT1 genes during grain development in *C. canephora*', *Plant Science*, 172, 978–96.

Liu, B., Falkenstein-Paul, H., Schmidt, W. and Beerhues, L. (2003). 'Benzophenone synthase and chalcone synthase from *Hypericum androsaemum* cell cultures: cDNA cloning, functional expression, and site-directed mutagenesis of two polyketide synthases', *The Plant Journal: for Cell and Molecular Biology*, 34, 847–55.

Long, M. C., Nagegowda, D. A., Kaminaga, Y., Ki Ho, K., Kish, C. M., Schnepp, J., Sherman, D., Weiner, H., Rhodes, D. and Dudareva, N. (2009). 'Involvement of snapdragon benzaldehyde dehydrogenase in benzoic acid biosynthesis'. *The Plant Journal: For Cell and Molecular Biology*, 59, 256–65.

Mahesh, V., Million-Rousseau, R., Ullmann, P., Chabrillange, N., Bustamante, J., Mondolot, L., Morant, M., Noirot, M., Hamon, S., de Kochko, A., Werck-Reichhart, D. and Campa, C. (2007). 'Functional characterization of two p-coumaroyl ester 3'-hydroxylase genes from coffee tree: evidence of a candidate for chlorogenic acid biosynthesis', *Plant Molecular Biology*, 64, 145–59.

Manach, C., Scalbert, A., Morand, C., Remesy, C. and Jimenez, L. (2004). 'Polyphenols: food sources and bioavailability', *American Journal of Clinical Nutrition*, 79, 727–47.

Márquez, L., García-Bueno, B., Madrigal, J. L. and Leza, J. C. (2012). 'Mangiferin decreases inflammation and oxidative damage in rat brain after stress', *European Journal of Nutrition*, 51 (6), 729–39. doi:10.1007/s00394-011-0252-x.

Martins, S. C. V., Araújo, W. L., Tohge, T., Fernie, A. R. and DaMatta, F. M. (2014), 'In high-light-acclimated coffee plants the metabolic machinery is adjusted to avoid oxidative stress rather than to benefit from extra light enhancement in photosynthetic yield', *PLoS ONE*, 9(4), e94862. doi:10.1371/journal.pone.0094862.

Matute, A. C., Sanchez Gomez, M. V., Campos, E. R., Alberdi, A. E., Gottlieb, M., Ibarretxe Bilbao, G., Delgado García, J. M., Gruart, I. Massó, A. and Leal Campanario, R. (2007). 'Compounds having neuroprotective properties', *WO* 2007077279 A1; ES2277567-A1 Spanish patent.

Mendoza-Sarmiento, G., Rojas-Hernández, A., Galano A. and Gutiérrez, A. (2016). 'A combined experimental–theoretical study of the acid–base behavior of mangiferin: implications for its antioxidant activity', *RSC Advances*, 6, 51171–82. doi:10.1039/C6RA06328D.

Miura, T., Ichiki, H., Hashimoto, I., Iwamoto, N., Kato, M., Kubo, M., Ishihara, E., Komatsu, Y., Okada, M., Ishida, T. and Tanigawa, K. (2001). 'Antidiabetic activity of a xanthone compound, mangiferin', *Phytomedicine*, 8 (2), 85–7.

Moerkercke, Van A., Schauvinhold, I., Pichersky, E., Haring, M. A. and Schuurink, R. C. (2009). 'A plant thiolase involved in benzoic acid biosynthesis and volatile benzenoid production', *The Plant Journal*, 60, 292–302.

Mondolot, L., La Fisca, P., Buatois, B., Talansier, E., de Kochko, A. and Campa, C. (2006). 'Evolution in caffeoylquinic acid content and histolocalization during *Coffea canephora* leaf development', *Annals of Botany*, 98 (1), 33–40.

Mösli Waldhauser, S. S., Kretschmar, J. A. and Baumann, T. W. (1997). 'N-methyltransferase activities in caffeine biosynthesis: biochemical characterization and time course during leaf development of *Coffea arabica*', *Phytochemistry*, 44 (5), 853–9.

Mouradov, A. and Spangenberg, G. (2014). 'Flavonoids: a metabolic network mediating plants adaptation to their real estate', *Frontiers in Plant Science*, 5, doi: 10.3389/fpls.2014.00620.

Moustafa, K. (2014). 'Improving plant stress tolerance: potential applications of engineered MAPK cascades', *Trends in Biotechnology*, 32 (8), 389–90. doi:10.1016/j.tibtech.2014.06.005.

Moustafa, K., AbuQamar, S., Jarrar, M., Al-Rajab, A. J. and Trémouillaux-Guiller, J. (2014). 'MAPK cascades and major abiotic stresses'. *Plant Cell Reports*, 33 (8), 1217–25. doi:10.1007/s00299-014-1629-0.

Murthy, P. S., Manjunatha, M. R., Sulochannama, G. and Madhava Naidu, M. (2012). 'Extraction, characterization and bioactivity of coffee anthocyanins', *European Journal of Biological Science*, 4, 13–19.

Muruganandan, S., Srinivasan, K., Gupta, S., Gupta, P. K. and Laj, J. (2005). 'Effect of mangiferin on hyperglycemia and atherogenicity in streptozotocin diabetic rats', *Journal of Ethnopharmacology*, 97, 497–501.

Mussatto, S., Ballesteros, L. F., Martins, S. and Teixeira, J. A. (2011). 'Extraction of antioxidant phenolic compounds from spent coffee grounds', *Separation and Purification Technology*, 83, 173–9.

Mussatto, S. (2015). 'Generating biomedical polyphenolic compounds from spent coffee or silverskin'. In V. R. Preedy (ed.), *Coffee in Health and Disease Prevention*, Elsevier, Department of Nutrition and Dietetics, King's College London, London, UK, pp. 93–105.

Nazario, G. M. and Lovatt, C. J. (1993). 'Separate *de novo* and salvage purine pools are involved in the biosynthesis of theobromine but not caffeine in leaves of *Coffea arabica* L.', *Plant Physiology*, 103 (4), 1203–10.

Niggeweg, R., Michael, A. J. and Martin, C. (2004). 'Engineering plants with increased levels of the antioxidant chlorogenic acid', *Nature Biotechnology*, 22 (6), 746–54.

Orians, C. M. (2000). 'The effects of hybridization in plants on secondary chemistry: implication for the ecology and evolution of plant-herbivore interactions', *American Journal of Botany*, 87, 1749–56.

Pal, P. B., Sinha, K. and Sil, P. C. (2014). 'Mangiferin attenuates diabetic nephropathy by inhibiting oxidative stress mediated signaling cascade, TNFα related and mitochondrial dependent apoptotic pathways in streptozotocin-induced diabetic rats', *PLoS ONE*, 9 (9), e107220. doi: 0.1371/journal.pone.0107220.

Panusa, A., Zuorro, A., Lavecchia, R., Marrosu, G. and Petrucci, R. (2013). 'Recovery of natural antioxidants from spent coffee grounds', *Journal of Agricultural and Food Chemistry*, 61, 4162–8.

Pinheiro, H. A., DaMatta, F. M., Chaves, A. R. M., Fontes, E. P. B. and Loureiro, M. E. (2004). 'Drought tolerance in relation to protection against oxidative stress in clones of *Coffea canephora* subjected to long-term drought', *Plant Science*, 167, 1307–14.

Pinto, M. M. M., Sousa, M. E. and Nascimento, M. S. J. (2005). 'Xanthones derivatives: new insights in biological activities', *Current Medicinal Chemistry*, 12, 2517–38.

Poiroux-Gonord F., Santini, J., Fanciullino, A-L., Lopez-Lauri F., Giannettini, J., Sallanon, H., Berti, L. and Urban, L. (2013). 'Metabolism in orange fruits is driven by photooxidative stress in the leaves', *Physiologia Plantarum*, 149, 175–87.

Pompelli, M. F., Martins, S. C., Antunes, W. C., Chaves, A. R. and DaMatta, F. M. (2010). 'Photosynthesis and photoprotection in coffee leaves is affected by nitrogen and light availabilities in winter conditions', *Journal of Plant Physiology*, 167 (13), 1052–60.

Prata, E. R. B. A. and Oliveira, L. S. (2007). 'Fresh coffee husks as potential source of antocyanins', *Food Science and Technology*, 40, 1555–60.

Procuraduría General de la República de Nicaragua Los Tuneros: Un movimiento trasformado con solidaridad y participación avanzando hacia el futuro, '*Procuraduría General de la República de Nicaragua.*'http://www.pgr.gob.ni/index.php/articulos/54-noticias/937-los-tuneros-un-movimiento-trasformado-con-solidaridad-y-participacivanzando-hacia-el-futuro.

Rakotomalala, J. J. R. (1992). 'Diversité biochimique des caféiers: analyse des acides hydroxy-cinnamiques, bases puriques et diterpènes glycosidiques. Particularités des caféiers sauvages de la région malgache (Mascarocoffea Chev.)', PHD thesis, 215pp.

Ramalho, J. C., Pons, T. L., Groeneveld, H. W. and Nunes, M. A. (1997). 'Photosynthetic responses of Coffea arabica leaves to a short-term high light exposure in relation to N availability', *Physiologia Plantarum*, 101, 229–39.

Ramalho, J. C., Pons, T. L., Groeneveld, H. W., Azinheira, H. G. and Nunes, M. A. (2000). 'Photosynthetic acclimation to high light conditions in mature leaves of Coffea arabica L.: role of xanthophylls, quenching mechanisms and nitrogen nutrition', *Australian Journal of Plant Physiology*, 27 (1), 43–51.

Ribereau-Gayon, P. (1968). *Les composés phénoliques des végétaux*. Dunod Paris.

Rivera, D. G., Balmaseda, I. H., León, A. A., Hernández, B. C., Montiel, L. M., Garrido, G. G., Cuzzocrea, S. and Hernández, R. D. (2006). 'Anti-allergic properties of *Mangifera indica* L. extract (Vimang) and contribution of its glucosylxanthone mangiferin', *Journal of Pharmacy and Pharmacology*, 58 (3), 385–92.

Rivera, D. G., Hernández, I., Merino, N., Luque, Y., Álvarez, A., Martín, Y., Amador, A., Nuevas, L. and Delgado, R. (2011). '*Mangifera indica* L. extract (Vimang) and mangiferin reduce the airway inflammation and Th2 cytokines in murine model of allergic asthma', *Journal of Pharmacy and Pharmacology*, 63 (10), 1336–45. doi:10.1111/j.2042-7158.2011.01328.x.

Rodríguez-López, N. F., Cavatte, P. C., Silva, P. E., Martins, S. C., Morais, L. E., Medina, E. F. and DaMatta, F. M. (2013). 'Physiological and biochemical abilities of robusta coffee leaves for acclimation to cope with temporal changes in light availability', *Plant Physiology*, 149 (1), 45–55.

Ross, I. A. (2005). '4*Coffea arabica*', In Medicinal plants of the world, Vol 3. *Chemical Constituents, Traditional and Modern Medicinal Uses*, Humana Press Inc., Totowa, NJ, pp. 155–96.

Rozo, Y., Escobar, C., Gaitán, Á. and Cristancho, M. (2012), 'Aggressiveness and genetic diversity of *Hemileia vastatrix* during an epidemic in Colombia', *Journal of Phytopathology*, 160, 732–40.

Schulthess, B. H., Ruedi, P. and Baumann, T. W. (1993). '7-glucosyladenine, a new adenine metabolite in coffee cell suspension cultures', *Colloque Scientifique International sur le Café* [C. R.], 15 (2), 770–2.

Schuster, B. and Herrmann, K. (1985). 'Hydroxybenzoic and hydroxycinnamic acid derivatives in soft fruits', *Phytochemistry*, 24, 2761–4.

Shahidi, F. and Chandrasekara, A. (2010). 'Hydroxycinnamates and their *in vitro* and *in vivo* antioxidant activities', *Phytochemistry Reviews*, 9, 147–70.

Shiroya, M. and Hattori, S. (1955). 'Studies on the browning and blackening of plant tissues. III. Occurrence in the leaves of Dahlia and several other plants of chlorogenic acid as the principal browning agent', *Plant Physiology*, 8, 358–69.

Shoji, K., Tsubaki, M., Yamazoe, Y., Satou, T., Itoh, T., Kidera, Y., Tanimori, Y., Yanae, M., Matsuda, H., Taga, A., Nakamura, H. and Nishida, S. (2011). 'Mangiferin induces apoptosis by suppressing Bcl-xL and XIAP expressions and nuclear entry of NF-κB in HL-60 cells', *Archives of Pharmacal Research*, 34 (3), 469–75. doi:10.1007/s12272-011-0316-8.

Silva, M. C., Nicole, M., Guerra-Guimaraes, L. and Rodriguez Jr, C. J. (2002). 'Hypersensitive cell death and post-haustorial defense responses arrest the orange rust (*Hemileia vastatrix*) growth in resistant coffee leaves', *Physiological Plant Pathology*, 60, 169–83.

Silva, M. C., Várzea, V., Guerra-Guimarães, L., Azinheira, H. G., Fernandez, D., Petitot, A-S., Bertrand, B., Lashermes, P. and Nicole, M. (2006). 'Coffee resistance to the main diseases: leaf rust and coffee berry disease', *Brazilian Journal of Plant Physiology*, 18 (1), 119–47.

Simmonds, P. L. (1864). *Coffee and Chicory: Their Culture, Chemical Composition, Preparation for Market, and Consumption*, London, E & FN Spon.

Sondheimer, E. (1958). 'On the distribution of caffeic acid and the chlorogenic acid isomers in plants', *Archives of Biochemistry and Biophysics*, 74, 131–8.

Suchal, K., Malik, S., Gamad, N., Malhotra, R. K., Goyal, S. N., Ojha, S., Kumari, S., Bhatia, J. and Arya, D. S. (2016). 'Mangiferin protect myocardial insults through modulation of MAPK/TGF-β pathways', *European Journal of Pharmacology*, 776, 34–43. doi:10.1016/j.ejphar.2016.02.055.

Takeda, T., Tsubaki, M., Kino, T., Yamagishi, M., Iida, M., Itoh, T., Imano, M., Tanabe, G., Muraoka, O., Satou, T. and Nishida, S. (2016). 'Mangiferin induces apoptosis in multiple myeloma cell lines by suppressing the activation of nuclear factor kappa B-inducing kinase', *Chemico-Biological Interactions*, 251, 26–33. doi:10.1016/j.cbi.2016.03.018.

Talamond, P., Mondolot, L., Gargadennec, A., de Kochko, A., Hamon, S., Fruchier, A. and Campa, C. (2008). 'First report on mangiferine (C-glucosyl-xanthone) isolated from leaves of a wild coffee plant, *Coffea pseudozanguebariae* (Rubiaceae)', *Acta Botanica Gallica*, 155, 513–19.

Treutter, D. (2006). 'Significance of flavonoids in plant resistance: a review', *Environmental Chemistry Letters*, 4, 147–57.

Trevisan, M. T. S., Farias de Almeida, R., Soto, G., De Melo E., Filho, V., Ulrich, C. M. and Owen, R. W. (2016). 'Quantitation by HPLC-UV of Mangiferin and Isomangiferin in Coffee (*Coffea arabica*) Leaves from Brazil and Costa Rica After Solvent Extraction and Infusion', *Food Analytical Methods*, 9, 1–7. doi:10.1007/s12161-016-0457-y

Tsubaki, M., Takeda, T., Kino, T., Itoh, T., Imano, M., Tanabe, G., Muraoka, O., Satou, T. and Nishida, S. (2015). 'Mangiferin suppresses CIA by suppressing the expression of TNF-α, IL-6, IL-1β, and RANKL through inhibiting the activation of NF-κB and ERK1/2', *American Journal of Translational Research*, 7 (8), 1371–81.

Wada, M. (2007). 'Foodstuffs compounding agent for treating diabetes comprises glycoside having xanthone structure', JP2007204462 - A Japanese patent.

Waller, G. R., Suzuki, T. and Roberts, M. F. (1980). 'Cell-free metabolism of caffeine in *Coffea arabica*', *Colloque Scientifique International sur le Café* [C. R.], 9, 627–35.
Waller, G. R., Jurzyste, M., Karns, T. K. B. and Geno, P. W. (1991). 'Isolation and characterization of ursolic acid from *Coffea arabica* L. (coffee) leaves', *Colloque Scientifique International sur le Café* [C. R.], 14, 245–7.
Weston, L. A. and Mathesius, U. (2013). 'Flavonoids: their structure, biosynthesis and role in the rhizosphere, including allelopathy', *Journal of Chemical Ecology*, 39, 283–97.
Winkel-Shirley B. (2001). 'Flavonoid biosynthesis. A colorful model for genetics, biochemistry, cell biology, and biotechnology', *Plant Physiology*, 126 (2), 485–93.
Yang, H., Duan, Y., Hu, F. and Liu J. (2004). 'Lack of altitudinal trends in phytochemical constituents of *Swertia franchetiana* (Gentianaceae)', *Biochemical Systematics and Ecology*, 32, 861–6.
Yang, H., Ding, C., Duan, Y. and Liu, J. (2005). 'Variation of active constituents of an important Tibet folk medicine *Swertia mussotii* Franch. (Gentianaceae) between artificially cultivated and naturally distributed', *Journal of Ethnopharmacology*, 98, 31–5.
Yang, Z., Weian, C., Susu, H. and Hanmin, W. (2016). 'Protective effects of mangiferin on cerebral ischemia-reperfusion injury and its mechanisms', *European Journal of Pharmacology*, 771, 145–51. doi:10.1016/j.ejphar.2015.12.003.
Yoosook, C., Bunyapraphatsara, N., Boonyakiat, Y. and Kantasuk, C. (2000). 'Anti-herpes simplex virus activities of crude water extracts of Thai medicinal plants', *Phytomedicine*, 6 (6), 411–9.
Zheng, X-Q., Nagai, C. and Ashihara, H. (2004). 'Pyridine nucleotide cycle and trigonelline (N-methylnicotinic acid) synthesis in developing leaves and fruits of *Coffea arabica*', *Physiologia Plantarum*, 122, 404–11.

Chapter 2

Nutritional and health effects of coffee

Adriana Farah, Federal University of Rio de Janeiro, Brazil

1 Introduction

Good health and well-being are essential for all humans and depend upon good nutrition. Only when an individual has good health can he or she fully utilize their physical and mental potentials.

Since the beginning of humanity, plant foods have been used to promote health and prevent disease. Coffee has been exalted by people of different nations and times not only because of its distinctive aroma and taste but also due to its stimulating and health-promoting effects (Bizzo et al., 2015).

The earliest potential references to coffee consumption are found in the Old Testament, where a bean was referred to as a 'parched pulse', and the first written mention of coffee is attributed to Razes, a tenth-century Arabian physician, who indicated that coffee cultivation may have begun as early as 575 AD (Smith, 1987; Folmer et al., 2017). However, the first written documentation of the medicinal properties and uses of coffee was reported by the Middle Eastern physician, Avicenna (980–1037 AD), who used it as a decongestant, muscle relaxant and diuretic infusion. It is said that in the thirteenth century, a doctor-priest from Mocha, Sheikh Omar, also discovered coffee in Arabia and used it as a cure for many different types of illnesses (Ukers, 1935). The earliest coffee houses opened in Mecca in the fifteenth century, but were primarily reserved for religious purposes. After they were popularized, following a trip to Aleppo, Dr. Leonard Rauwolf, a German physician, introduced the beverage to Western Europe in the sixteenth century and referred to it as being 'almost as black as ink and very good in illness, chiefly that of the stomach' (Ukers, 1935; Folmer et al., 2017).

http://dx.doi.org/10.19103/AS.2017.0022.14

Neither the Greek physician Hippocrates (460–377AD) nor the Roman physician Aelius Galenus (129–199 AD), who later expanded Galenic theories, ever mentioned coffee, as it was not consumed in their region at that time. However, the followers of Hippocratic–Galenic medicine, which dominated physiology from the fourth century BC to the nineteenth century, used it to balance the body's 'humours' in accordance with individual temperaments. Coffee was considered to be beneficial to people with a lymphatic or bilious temperament, whereas sanguine or nervous subjects were advised to use it more reservedly. Consequently, coffee houses, which were first introduced in Europe in the seventeenth century, were often recommended for those suffering from maladies as part of their treatment (Bizzo et al., 2015). In the eighteenth century, they also became places for social gathering and commerce, and over time, more coffee houses opened up and became popular (Folmer et al., 2017).

Despite its recognition as a medicinal agent or simply a beverage with an attractive taste, throughout history, coffee has occasionally been seen as a villain, and to date, remainders of this reputation still exist. The perception of coffee as an intoxicating drug and the sensitivity of some people to caffeine are the main reasons for this, along with past discussions on its potential contribution to the development of cancer or other diseases. However, over the last few decades, the appearance of modern scientific technology, in combination with large and reliable databases and sophisticated statistics, has enabled the separation of confounding factors in epidemiological studies such as existing medical conditions, smoking or a poor-quality diet. Additionally, an increasing number of studies have proved that despite its nutritional limitations, coffee is a complex mixture of bioactive substances that may act together to help prevent diseases when consumed appropriately. In view of this, our understanding of coffee and its healthful properties has changed dramatically. Currently, the general opinion is that moderate coffee consumption is not associated with increased long-term risks amongst healthy individuals and can be incorporated into a healthy and diverse diet, in combination with other healthful behaviours (US Department of Agriculture – USDA, 2015).

This chapter presents a brief literature review of the nutritional and main health-related aspects of regular coffee consumption.

2 Nutrients and bioactive compounds of coffee

The chemical composition of roasted coffee beans has been detailed in previous chapters. In summary, they contain approximately 43% carbohydrates (of which 70–85% is comprised of polysaccharides, arabinogalactans, mannans and glucan, and the remaining amount includes sucrose, reducing sugars, lignins and pectins); 7.5–10% proteins; other nitrogenous compounds (1% caffeine, 0.7–1% trigonelline and 0.01–0.04% nicotinic acid); 10–15% lipids (of which approximately 75% correspond to triacylglycerols, 18.5% to esters of diterpens and free diterpens and the remaining amount to esters of sterols, free sterols, sterylglucosides, waxes, tocopherols and phosphatides); 25% melanoidins, 3.7–5% minerals and ~6% organic and inorganic acids, and esters (1–4% chlorogenic acids and other phenolic compounds, 1.4–2.5% aliphatic acids and quinic acid and <0.3% inorganic acids), in addition to other minor compounds that may be exclusive to a particular species

(Farah, 2012; Speer and Kölling-Speer, 2017). In this chapter, we will focus on the most common species found around the world, *Coffea arabica* and *Coffea canephora*, which are also the most consumed and the only species studied with respect to human health.

The coffee beverage or brew is an aqueous extract derived from the infusion or percolation of roasted and ground coffee beans, using hot or cold water. The reported amounts of nutrients, main bioactive compounds and other non-nutrients in coffee brews are presented in Table 1. It is worth noting that a coffee brew exhibits very high variability in terms of chemical composition due to many possible variations in raw material production, processing and brewing, which lead to the final product (the brew). The first variable is the blend, which can contain different percentages of coffee beans, each with distinct chemical compositions derived from genetic aspects, origin and degree of maturation (the latter being considered only in the case of a lower quality blend), grown under different conditions and processed via varying postharvest methods. There are also many existing types of roasting profiles and roasting degrees. The roasted beans can then be ground to different sizes and the proportion of powder to water classically used can change dramatically between countries and cultures. For example, whilst in most European countries, the use of 6 g per 100 mL is common for filtered coffees, in Brazil, 10 g or more is used. In Italy, 20 g of ground roasted coffee is also not uncommon for 100 mL. In espresso coffee, although traditionally 6–8 g is used per 25 mL water, an extreme proportion of 10 g per 25 mL water is nowadays often used by third-wave baristas. In addition, there are a variety of brewing methods where pressure, temperature and contact time between the water and the ground coffee vary, and therefore require different amounts of powder. Some methods use filters made up of different materials, which may also influence the composition of the final brew. On top of all this, the size of a cup can vary from about 25 mL for an Italian espresso to 600 mL (20 oz) in the United States. The standard American cup, however, is often reported as being equivalent to approximately 250 mL (8 fluid oz). The traditional European filter cup has been defined in different studies, including that of Floegel et al. (2012), as containing 150 mL. Finally, analytical methods may cause differences in the reported compositional results, especially for the least sensitive and specific methods. Thus, it is understandable that there are no rigorous standard values that represent the chemical composition of a cup of coffee, but rather a range of values. Table 1, therefore, reports the ranges of values found in the literature, but higher or lower values can be found, depending on the compound.

It is worth noting that the yield of a brew (i.e. the amount of solids extracted from the ground coffee found in a cup) may vary from as low as 14% to as much as 60% during soluble coffee production, thus affecting the composition of the brew. Further to this, the extractability of a compound by water will also depend on the amount of soluble solids in the water. Water, containing a high amount of minerals like calcium, magnesium and chloride may extract less solids from the ground coffee, and may also influence the flow time in an espresso machine (Folmer, pers. comm.). Considering that up to 40 g of coffee could be used to prepare 100 mL of a coffee beverage, the amount of soluble solids in coffee brews has been reported to vary from 2 to 6 g per 100 mL cup (Pettraco, 2005) (also referred to as a TC ranging from 2 to 6%). In the traditionally weaker American coffees, measurements of the amount of solids in a few cups yielded 1.2–1.5 g per 100 g. It is worth highlighting that the amount of solids depends upon the degree of roasting.

The main components of the brew are now briefly described.

Table 1 Content range of nutrients, bioactive compounds and other non-nutrients in coffee brews obtained from ground and roasted blends of *C. arabica* species or blends of *C. arabica and C. canephora* species (Trugo, 1985; Macrae, 1985; Clifford, 1985; Urgent et al., 1995; Nunes et al., 1997; Balzer, 2001; Alcázar et al., 2003; Petracco, 2005; Boekschoten et al., 2006; Lang et al., 2010,2011; Farah, 2012; Rubach et al., 2014; Lachenmeier, 2015; USDA, 2017; Glória and Engeseth, 2017)

Nutrients and non-nutrients	Content range[a] (from blends of *C. arabica* or *C. arabica* and *C. canephora* sp.)
Macronutrients	*mg per 100 mL*
Water	94 000–98 500 (TC 1.5–6%)
Simple sugars[b]	0–100 (one report up to 200 mg)
Proteins	120–400
Lipids	180–400
Soluble fibres[c]	200–700 (more commonly between 400 and 500)
Aliphatic acids and quinic acid[f]	692–2140
Vitamins:	
Thiamin (B1)	0.001
Riboflavin (B2)	0.177
Niacin (B3, nicotinic acid)[d]	0.8–10 (more commonly up to 5)
Pyridoxine (B6)	0.002
Folate (B9, DFE)[e]	1
Vitamin C, total ascorbic acid	0.2
Vitamin E (alpha-tocopherol)	0.01
Vitamin K (phylloquinone)	0.1
Tocopherols (α, β, γ)	Traces, only in unfiltered coffees
Minerals: Total ashes	150–500
Potassium, K	115–320
Calcium, Ca	2–4
Sodium, Na	1–14
Phosphorus, P	3–7
Iron, Fe	0.02–0.13
Zinc, Zn	0.01–0.05
Manganese, Mn	0.02–0.05
Bioactive compounds	*mg per 100 mL*
Caffeine	50–380 (commonly between 50 and 150)
Trigonelline	12–50
N-methylpyridinium	2.9–8.7
Diterpenes (cafestol and kahweol)	0.2–1.5 (paper filtered); 2.6–10 (boiled)

Nutrients and non-nutrients	Content range[a] (from blends of *C. arabica* or *C. arabica* and *C. canephora* sp.)
Chlorogenic acids	32–500 (commonly 50–150)
Sum of other phenolic compounds	0.1–0.2
Melanoidins	500–1500
β-carbolins (norharman and harman)	0.004–0.08
Serotonin	0–1.4
Melatonin	0.006–0.008
Polyamines (spermine and spermidine)	0.4
Some undesirable compounds[g]	*µg per 100 mL*
Acrylamide	3.9–7.7
5-hydroxytryptamides[h]	1.2–34.3 (filtered), 350–840 (espresso and French press)
Furan[i]	3.8–262

[a] Content varies with blend, origin, agricultural practices, roasting method and degree, grid, brewing method, amount of coffee powder to water and analytical method.
[b] Arabinose, mannose, galactose, sucrose and minor monosaccharides.
[c] Polysaccharides, mainly galactomannans and type II arabinogalactans.
[d] Daily recommendations for adults:16 mg for men and 14 mg for women (WHO/FAO, 2002).
[e] Dietary Folate Equivalent.
[f] pH 4.3 (acidic coffee, light roast), to 5.8. Common values around 5.0.
[g] These undesirable compounds do not include incidental contaminants.
[h] N-alkanoyl-5-hydroxytryptamides (C5HTs).
[i] Content decreases rapidly after brew preparation.

2.1 Macronutrients

As with other types of plant beverages, coffee brews do not contain excessive amounts of macronutrients (absorbable carbohydrates, proteins and lipids) and calories, unless they are consumed with sugar, milk or cream. According to the USDA National Nutrient Database (2017), 100 mL of filtered coffee (breakfast blend) contains approximately 2 kcal. Exceptions include boiled coffees and similar types of brews, which contain a reasonable amount of lipids as both soluble and insoluble materials are consumed. The nutritional quality of coffee proteins is limited and approximately 50% of this fraction is insoluble and lacks the essential amino acid, tryptophan (Macrae, 1985). In addition, during roasting, most simple sugars and proteins are degraded or changed via the Maillard and pyrolysis reactions and so the amount of protein in the cup is low (120–400 mg per 100 mL, USDA 2017). The detection of up to 200 mg per 100 mL of simple sugars, mainly arabinose, galactose and mannose, and lower amounts of sucrose, fructose and glucose, has been reported for brewed coffee and dissolved soluble coffee at 2% (coffee/water), with the latter being higher (Macrae, 1985; Trugo, 1985). Galactomannans and type 2 arabinogalactans are considered to be the predominant soluble dietary fibres in a coffee beverage (often between 140 and 650 mg per 100 mL), of which approximately 70% is comprised of galactomannans (Gniechwitz et al., 2007; Farah, 2012); however, care is needed to distinguish polysaccharides from total carbohydrates in order to avoid overestimation.

2.2 Micronutrients

With regards to micronutrients, the brew may contain a reasonable amount of vitamins and minerals. Niacin, in the form of nicotinic acid, is the main vitamin in a coffee brew and is also known as vitamin B3 or PP (pellagra preventing), and participates as a coenzyme in various metabolic processes involved in energy metabolism and tissue health. Additionally, it contributes to the normal functioning of the nervous system, among many other functions. Deficiency of this vitamin causes pellagra, a disease characterized by skin lesions and diarrhea, among other symptoms (Arauz et al., 2015). Regular coffee consumption can supply an essential part of the daily recommendation for adults, which is 16 mg for men and 14 mg for women according to FAO standards (WHO/FAO, 2002). Niacin can also be produced in the liver from a reasonable amount of tryptophan (60mg thyptophan makes 1mg niacin), which is present in animal proteins, nuts, and some other seeds. The intake of niacin from coffee is particularly important in places where tryptophan consumption is low, such as in the rural and less developed areas of Central America and Central Africa where corn (which is poor in niacin and tryptophan) is the staple food (Carpenter, 1983; Macrae, 1985). However, it is worth noting that people who consume corn as tortillas or similar foods made with corn flour pre-treated with alkaline water are not at risk of niacin deficiency as those who consume untreated corn and corn flour (Carpenter, 1983). Most recently, supplementation of nicotinic acid has been used to decrease low and very low density lipoproteins (LDL and VLDL) levels (Le Bloch et al., 2010) and to contribute to the hepatoprotective effect of coffee (Arauz et al., 2015) and therefore, an additional food source of niacin, like coffee is welcome considering that excess amounts of the vitamin are excreted in urine (Wang et al, 2001).

Coffee brews may also contain very small amounts of nicotinamide, another form of niacin, and other B vitamins (thiamin, riboflavin, pyridoxine, folic acid), ascorbic acid (vitamin C), and phylloquinone (vitamin K) (Macrae, 1985; USDA 2017). Unfiltered coffees can contain small residual amounts of tocophenols (α, β and γ – the latter two predominate), although during and after roasting almost all of the original amount in green coffee (about 60 mg/100g) is degraded or oxidized (Macrae, 1985).

Considering the different factors that create a large variability in the composition of roasted coffee, in addition to differing brewing methods and the extractability of different minerals, the values of total ash ranging between 150 and 500 mg per 100 mL have been reported, including data from dissolved soluble coffee (containing 7–10% ash), prepared at 2% (ground coffee to water). For infusions prepared from ground coffee in Poland (Grembecka et al., 2007), the consumption of 300 mL (2 cups) was estimated to supply, on average, 4.5% of the recommended dietary allowance (RDA) for Mg, 3.5% for K, 2.8% for Mn, 2.4% for Cr, 1.9% for P, 0.32–0.43% for Ca and Na, 0.26–0.33% for Cu and Fe, 0.13% for Zn and 2.6–15.6% for Ni. Higher average percentages were estimated from the ingestion of a 300-mL beverage prepared from instant coffee: 12.3% for Mg, 8.9% for K, 8.6% for Mn, 4.9% for Cr, 7.4% for P, 1.6% for Ca, 2.5% for Na, 0.32% for Cu, 2.9% for Fe, 0.35% for Zn and 4.9–29.7% for Ni. Estimates indicate that the consumption of coffee in typical amounts does not exceed the tolerance limits for the ingestion of toxic metals, such as Pb and Cd.

In a study performed in Portugal (Oliveira et al., 2012), two cups of instant coffee (total of 4 g) were estimated to supply 9.5% of the RDA for K, 5.2% for Mg, 4.4% for Mn, 3.5% for Ni, 2.2% for P, 1.5% for Fe, 0.5% for Cr, 0.4% for Ca and 0.2% for Na. In a more recent study, also performed in Portugal (Oliveira et al., 2015), one cup of an espresso coffee beverage (prepared from 5 to 6 g of ground coffee) was estimated to provide 5.2–7.0% of the RDA for

K, 2.8–7.2% for P, 1.4–2.2% for Mg, 1.4–1.9% for Mn, 0.14–0.28% for Ca, 0.07–0.15% for Fe and <0.02% for Na.

Other similar studies indicate that regular coffee intake (300 mL) does not contribute, in most cases, to amounts higher than 10% of the daily recommendations in different countries.

2.3 Bioactive compounds

It is well known that nutrients are required to maintain normal body functions and are therefore bioactive. However, the term 'bioactive compounds' commonly refers to minor food constituents that exert biological functions other than nutritional functions. These compounds are commonly found in plants, and in a few animals that feed on them, and their chemical structures and biological functions vary widely.

Although most health-related aspects of coffee have been attributed to the beverage and not to individual compounds, mechanistic studies have suggested that some specific bioactive compounds play key roles as co-adjuvant agents in disease prevention. In addition to caffeine, the most studied bioactive compounds of coffee are chlorogenic acids and their lactones, trigonelline and their derivatives, the diterpenes, cafestol and kahweol and melanoidins. The polysaccharides, galactomannans and type II arabinogalactans, and β-carbolines are amongst the emerging bioactive compounds for which there is still insufficient information to substantiate any health effects. Also, relatively recently, it has been suggested that some coffee amines exert positive effects on health when consumed in low quantities, which are referred to as bioactive amines. Each of these compounds or group of compounds will be introduced hereafter and their effect on health will be briefly discussed. The main undesirable compounds in coffee are also introduced.

2.3.1 Caffeine

Caffeine is the most well-studied compound present in coffee, and its mechanisms of function are generally well documented. First, there are psychostimulating effects which include an acute impact on mental performance as well as long-term influence on the risk of developing neurodegenerative diseases such as Parkinson's and Alzheimer's. Caffeine has also been found to improve physical performance, which will be discussed in the next section on the health effects of coffee. Most recently, a number of studies have reported new bioactive effects for caffeine, and one of the emerging effects is the enhancement of the antioxidant effect of coffee. Caffeine metabolites, especially 1-methylxantine and 1-methylurate, have exhibited an antioxidant activity *in vitro* (Moura-Nunes et al., 2009). Corroborating these results, the average plasma iron-reducing capacity of human subjects after regular coffee consumption was higher than that recorded after the consumption of decaffeinated coffee, suggesting that whole coffee is more efficient than decaffeinated coffee with respect to its antioxidant capacity (Moura-Nunes et al., 2009).

There are a few *in vitro* studies showing that caffeine contributes to the antibacterial effect of coffee against *Streptococcus mutans*, a significant contributor to cariogenic bacteria, as well as intestinal pathogenic bacteria (Antonio et al., 2010). Over the last few years, it has also been suggested that caffeine exerts an antihyperlipidemic effect (decreased storage of triglycerides and cholesterol) by inhibiting lipogenesis and stimulating lipolysis through the regulation of the gene expression responsible for lipid metabolism in liver cells (Quan, 2013). These are just a few of the various emerging effects of caffeine.

2.3.2 Chlorogenic acids and their derivatives

Chlorogenic acids and their derivatives – Chlorogenic acids are the main phenolic compounds in coffee and this group of compounds includes approximately ten major esters and four major lactones (produced during roasting), in addition to dozens of trace compounds. The total amount of chlorogenic acids in *C. canephora* beans is almost double than that found in *C. arabica* beans, and because chlorogenic acids are partly degraded or transformed during roasting, dark roasted coffees contain lower amounts of these compounds. In roasted products, the difference between species is significantly reduced.

These compounds are frequently referred to as powerful antioxidants and anti-inflammatory compounds due to the results of *in vitro* and animal studies, as well as a few human studies (Santos et al., 2006; Torres and Farah; 2016, Folmer et al., 2017), but the mechanism of action of the different compounds and how they are related to the prevention of the disease remains, to a large extent, unknown.

Owing to the high concentration of chlorogenic acids in coffee brews (Table 1), compared with chlorogenic acids and other phenolic compounds in foods, in general, they may play a major role in the diet of consumers as a source of antioxidative compounds. The significant contribution of chlorogenic acids to the dietary intake of antioxidative compounds is exemplified in a number of reports from different countries in which, based on their official food consumption database or other types of surveys, coffee was the main contributor to total dietary antioxidant capacity, that is, Brazil (66%), Norway (64%), Italy (38% for women and 27% for men), Spain (45%), Japan (56%) and the Czech Republic (54.6% for women and 43.1% for men) (Torres and Farah, 2016). However, it should be kept in mind that the intake of chlorogenic acids from coffee does not replace the intake of antioxidants from other foods, as each has its own specific bioactivity.

Together with other polyphenols, carotenoids and additional classes of antioxidative compounds, chlorogenic acids and their lactones have been associated with a decrease in the risk of Alzheimer's and type 2 diabetes, amongst various degenerative diseases (Kasai et al., 2000; Obuleso et al., 2011; Farah, 2012).

Long before epidemiological studies investigated the association between coffee consumption and health effects, the antimutagenic property of chlorogenic acids and their metabolites was discovered. Recent studies have confirmed these findings and elucidated several mechanisms of chemopreventive action, which include modulating the expression of the enzymes that are involved in endogenous antioxidant defences, DNA replication, cell differentiation and ageing (Feng et al., 2005; Ramos, 2008;Jurkowska, 2011), metal chelation, inactivation of reactive compounds and metabolic pathway changes (Kasai et al., 2000; Farah, 2012). In the colon, for example, chlorogenic acids may inactivate free reactive radicals and as a result help prevent colon cancer (Ludwig et al., 2014).

Additional health effects observed *in vitro* and in animal studies include hepatoprotective (including cirrhosis, liver cancer and other liver diseases), immune-stimulatory and antibacterial and antiviral activities. Synthetic derivatives of these compounds have also inhibited HIV-1 replication in cells, which could play a role in research towards drugs that inhibit HIV (Farah, 2012). Additionally, recent *in vitro* studies have suggested that after coffee consumption, the unabsorbed portion of chlorogenic acids may serve as a substrate to stimulate the growth of beneficial intestinal bacteria; however, this effect requires further investigation due to conflicting data in different studies (Sales et al., 2017).

Trace amounts of other phenolic compounds, that is, isoflavones, proantocyanidins and lignans, have been identified in coffee (Farah and Donangelo, 2006), which possibly enhance the beneficial effects of chlorogenic acids.

2.3.3 Melanoidins and polysaccharides

Melanoidins and polysaccharides – Coffee melanoidins have gained importance over the years because of their contribution to, amongst others, the antioxidant and antimicrobial effects of coffee (Rufián-Henares and Morales, 2007). This is at least, in part, due to the incorporation of chlorogenic acids and other bioactive compounds into their structure during roasting (Farah, 2012).

As melanoidins are not digested, they may act, in combination with coffee polysaccharides (mainly galactomannans and type II arabinogalactans), as soluble dietary fibres. They are largely indigestible and thus fermented in the gut (Borrelli et al., 2004; Gniechwitz, 2008). A recent study concluded that the consumption of 0.5–2 g melanoidins per day (present in 2–5 cups) contributes up to 20% of the recommended 10 g of daily soluble dietary fibre intake. It has also been hypothesized that these substances may stimulate the growth of beneficial bacteria in the lower digestive tract (Fogliano and Morales, 2011), in the same way as chlorogenic acids; however, the data remain controversial.

As with chlorogenic acids (Passos et al., 2014), it has been hypothesized that melanoidins can enhance immune-stimulating properties and contribute significantly to reducing the risk of colon cancer (Vitaglione et al., 2012; Moreira et al., 2015; Fogliano and Morales, 2011), which might occur in different ways: i) by increasing the elimination rate of carcinogens through higher colon motility and faecal output, ii) by decreasing colon inflammation through improved microbiota balance (prebiotic effect) and iii) by serving as a 'sponge' for free radicals in the gut (Garsetti et al., 2000; Folmer et al., 2017).

2.3.4 Trigonelline and derivatives

Trigonelline and derivatives – Trigonelline is another compound that has gained importance in recent years due to its potential contribution to the protective effect of coffee against diseases. *In vitro* and animal studies have reported different involvements of trigonelline against type 2 diabetes (Yoshinari and Igarashi, 2010), as well as neuroprotective (Hong et al., 2008; Tohda et al., 2005), antitumour (Hirakawa et al., 2005) and phytoestrogenic effects (Farah, 2012). Beans from *C. arabica* species contain higher amounts of this compound, compared with *C. canephora*, and as with chlorogenic acids, trigonelline undergoes changes and degradation during roasting; hence, dark roasted coffees contain low amounts. However, 10–20% of the original amount of trigonelline is converted into nicotinic acid (niacin) (Farah, 2012). In addition to its vitamin function, niacin is also involved in other bioactive functions, presenting antidiabetic (Yoshinari and Igarashi, 2010), antioxidant and hepatoprotective (Arauz et al., 2015) effects *in vitro* and in animal studies. The compound *n*-methylpyridinium and other pyridinium derivatives are additional thermal degradation products generated by the decarboxylation of trigonelline. It has been reported that like trigonelline, *n*-methylpyridinium promoted higher glucose utilization in liver cells, stimulating cellular energy metabolism and contributing to the protective effect against type 2 diabetes (Riedel et al., 2014). Pyridinium derivatives have also been reported to present antioxidant/chemopreventive (Somoza et al., 2003), hepatoprotective (Gebicki et al., 2008), vasoprotective (Lang et al., 2011) and antithrombotic effects (Kalaska et al., 2014).

2.3.5 Diterpenes

Diterpenes – Cafestol and kahweol are diterpenes present in coffee mainly in the form of salts or esters of (predominantly) saturated and unsaturated fatty acids. They represent approximately 20% of the lipid fraction of coffee, with cafestol being more abundant. Higher levels of diterpenes are found in *C. arabica* than in *C. canephora* species. Coffee diterpenes exhibit strong anticarcinogenic and hepatoprotective properties *in vitro* (Farah, 2012).

Diterpene levels in the coffee cup vary significantly based on the natural variations in green coffee beans, roasting conditions and preparation methods (Urgert et al., 1997; Gross, 1997;Urgert and Katan, 1995). Whilst filtered and soluble coffees are practically diterpene-free (due to their poor solubility in water, they are trapped by paper filters), espresso-based methods contain higher levels of diterpenes, which are, on the other hand, significantly lower than those found in French press or Turkish coffee (2–10 mg per cup) (Farah, 2012).

2.3.6 β-carbolines

β-carbolines – These are alkaloids formed in coffee mainly during roasting and the two identified β-carbolines in coffee are norharman and harman. Despite some past controversies regarding the neurological and toxicological effects of these compounds in studies using high doses in animals, β-carbolines have been recently associated with potentially positive effects, including neurological ones, with antidepressive and neuroprotective properties. It has also been suggested that they may reduce the risk of diabetes. The total concentration of these compounds in the brew is highly variable in the literature, from 4 to 80 μg per 100 mL, but typical concentrations are reported to be in the range of 4–20 μg per 100 mL, being primarily dependent on the coffee species. Roasted *C. canephora* beans have consistently higher amounts of β-carbolines than *C. arabica* beans (Farah, 2012;Rodrigues and Casal, 2017;Casal, 2017).

2.3.7 Bioactive amines

Bioactive amines – These compounds are organic bases with psychoactive, neuroactive or vasoactive activity and participate in a number of processes in the human body. Coffee amines that present positive health effects in *in vitro* and in animal studies, other than their known physiological effects in the body, are called bioactive amines. However, no direct association between their presence in coffee and benefits to human health has been found. The main bioactive amines are the indolamines, serotonine and melatonin, and the polyamines, spermine and spermidine. Mean reported concentrations of serotonin in coffee beverages vary from non-detected to 90 μg per 100 mL in most coffee beverages, but from 372 to 1354 μg per 100 mL in Turkish coffee. Information regarding melatonin is rare; however, levels ranging from 6 to about 8 μg per 100 mL have been found. Although serotonin is a neuroactive substance with various positive effects on well-being, serotonin from the diet cannot cross the blood–brain barrier and can only be produced in the brain; however, serotonin from the diet can have other potentially relevant roles including broncho- and vasoconstrictor, antihypertonic, antioxidant and antiallergic and antidiuretic effects. It can also help to modulate the volume and acidity of gastric juice. Spermine and spermidine (reported amounts for each vary from non-identified to more than 150 μg per 100 mL brew) are efficient free radical scavengers in several chemical and enzymatic systems. Other amines have also shown positive effects on health when in low quantities (Gloria and Engeseth, 2017).

2.4 Undesirable compounds in coffee

A few compounds derived from microbial contamination (ocratoxin A, biogenic amines), pesticides or chemical reactions that occur during the roasting process (mainly acrylamide, furan and polycyclic aromatic hydrocarbons – PAH) have been of concern to health authorities such as the Food and Drug Administration (FDA), Food and Agriculture Organization of the United Nations (FAO) and the European Food Safety Authority (EFSA).

2.4.1 Ocratoxin A

Ocratoxin A and similar toxins are derived from green bean contamination with mould and can be avoided by carefully harvesting, processing and storing coffee, which is reflected in good quality. Ocratoxin A is gradually degraded by the high temperatures of roasting and its residual levels in roasted coffee are regulated in many countries. In the European Union, regulation 1881/2006 states that for roasted coffees, the maximum limit is 5 μg per kg.

2.4.2 Pesticides

Pesticides comprise a large number of substances belonging to different chemical groups, which are used to control plant diseases, pests or weeds. They can be neurotoxic or inhibit vital metabolic reactions in living beings, targeting different mechanisms, and individual pesticides present different levels of toxicity. In order to protect human safety and health resulting from pesticide application during coffee production, many countries have put these chemicals under strict legislation and surveillance. For example, in the United States, the FDA establishes the maximum amount of a pesticide allowed to remain in food, as part of the process of regulating pesticides. Presently, tolerance limits for about 43 pesticides in coffee are listed by the FDA. In Japan, the maximum tolerated residual limits are amongst the lowest (often 0.01 ppm). Despite all the regulations, pesticide residues have been found via analyses of green coffee performed prior to importing at differing occasions and in different countries. Due to restrictions and monitoring, the residual levels of pesticides in commercial ground and roasted coffees are usually very low and within the established limits. The solubility of residual pesticides after roasting is often low and therefore low amounts are found in the brew, especially in filtered coffee. However, the toxicity of their metabolites in seeds or degradation products during roasting has not been well studied and could be higher than that of the pesticides themselves (Farah, 2012; Cunha and Fernandes, 2017).

2.4.3 Acrylamide, furan and PAH

Acrylamide, furan and PAH are derived from reactions that occur during roasting, more specifically Maillard and pyrolysis (Farah, 2012), which can occur in many other heated foods such as French fries or bread. Studies assessing the risk of their concentrations in brews have not found considerable amounts which could cause harm to human health, as epidemiological studies have failed to find a link between these compounds in coffee and an elevated risk of cancer or other diseases (Lipworth et al., 2012; Nkondjock, 2012).Amongst all these compounds, acrylamide is the most abundant and the one which many food and health authorities have been most concerned about. It is formed at the beginning of the roasting process and its levels decrease thereafter to a certain degree (Farah, 2012). Acrylamide has been associated with cancer in one study using laboratory rodents which were exposed to extremely high concentrations (1000–10 000 times physiological ranges) (Mucci and Adami, 2009). Whilst the US FDA (FDA, 2016) reported that coffee is a significant source of acrylamide exposure for adults, the EFSA's recent opinion on acrylamide stated that health authorities do not see any direct evidence of cancer risk (EFSA 4104, 2015b; Folmer et al., 2017). The

European Commission's recommended indicative value is 450 μg acrylamide per kg of roast and ground coffee, a level which is generally achievable for commercial products.

Furan, hydroxymethylfurfural (HMF) and furfural are heterocyclic, low molecular weight molecules with a furan ring and potential carcinogenicity in common. Furan has been classified as a possible carcinogen (Group 2B) by the International Agency for Research on Cancer (IARC). Group 2B status is assigned to compounds and exposure conditions for which there is limited evidence of carcinogenicity in humans and insufficient evidence of carcinogenicity in experimental animals, and therefore requires further investigation. No human studies are available regarding the effects of furan and there is a significant uncertainty in the extrapolation of risk from animal assays performed in the laboratory to the equivalent risk for humans. When dealing with furfuryl and furan derivatives, the EFSA report (EFSA, 2011) concluded that notwithstanding some indications of *in vitro* genotoxicity, based on available data of exposure and on *in vivo* genotoxicity studies, which gave negative results for the carcinogenicity evaluation in rats and mice, under normal conditions, these compounds are of no concern to human health (Folmer et al., 2017). Regulatory bodies, including the US FDA and the EFSA, have made no recommendations regarding the maximum levels of furan in the dietary intake.

In coffee, these compounds are formed during the roasting stage mainly via thermal degradation/Maillard reaction of reducing sugars, alone or in combination with amino acids or via the thermal degradation of amino acids. The consumed levels of furan in coffee are highly variable and reflect not only the preparation methods but also the roasting conditions; however, these compounds are not exclusive to coffee. The major contributors to furan exposure in adults and teenagers were estimated to be fruit juice, and milk-based and cereal-based products. Additionally, jarred baby foods were also major contributors in toddlers (Ferreira et al., 2017).

Examples of a range of furan concentrations, obtained using coffees from the Spanish market prepared by different methods, are 12–146 μg per litre with the lowest values found in instant and filtered coffee and the highest values found in espresso. Boiled coffee was not evaluated. The concentration for commercially packed coffee capsules was approximately 240 μg per litre. The furan content of coffee brews from automatic coffee vending machines ranged from 11 to 262 μg per litre; however, this is not representative of what people would consume. Due to its high volatility, after coffee preparation, losses occur rapidly upon mixing and waiting for the brew to cool down (Ferreira et al., 2017).

The most important contributors of HMF in the diet are dried fruits, caramel, vinegar, bread and coffee. The content of HMF in coffee samples from coffee vending machines (8 g of ground coffee to 100 mL water) ranged between 4 and 60 mg per litre with a mean content of 28.8 mg per litre. The concentrations of furfural in coffee brews from European vending machines ranged between 0.30 and 1.30 mg per litre (Ferreira et al., 2017). To date, no measures have been identified to mitigate furan without impacting the typical coffee aroma.

PAH, from which benzo[a]pyrene is the most relevant from a toxicity point of view, can be formed in coffee and other foods that are severely roasted or exposed to very high temperatures. This compound is classified by the IARC as probably carcinogenic to humans. However, the level of exposure to PAH from coffee is low and within the safe limits set by International Agencies (IARC, 2010; EFSA, 2008).

2.4.4 Biogenic amines

Biogenic amines are organic bases of low molecular weight that participate in the regular metabolic processes of plants, microorganisms and animals. They are produced in the body and can also be provided by the diet and from the microbial flora of the intestine; at high concentrations, however, they can pose a toxicological risk.

Examples include histidine, tyramine, tryptamine, cadaverine and putrescine. Histidine is the most toxic and is associated with a hypotensive effect and headaches. Putrescine, cadaverine and tyramine seem to be toxic in higher doses in animals, but the individual sensitivity to these compounds in humans varies considerably and causes different responses. In coffee, biogenic amines originate from the action of microbial enzymes on amino acids during fermentative processes, suggesting inappropriate storage or low-quality defective fermented seeds. Roasting may deconjugate and increase the amount of some free biogenic amines, whilst most other amines are degraded (Gloria and Engeseth, 2017b; Farah, 2012).

3 Main beneficial health effects of coffee

The role of coffee drinking for the purpose of socialization and relaxation is indirectly important for health, as stress plays a major role in the development of several diseases that may lead to serious complications and death. Additionally, coffee has been shown to help prevent degenerative disorders, many of which are related to neurostimulating, antioxidant and anti-inflammatory effects. A prospective US cohort study (Freedman et al., 2012) examined the association of coffee drinking with subsequent cause-specific and total mortality in the National Institutes of Health– AARP Diet and Health Study. This study involved more than 400000 people and is, so far, the largest human study investigating coffee and health. A significant inverse association between coffee and specific deaths due to heart disease, respiratory disease, stroke, injuries and accidents, diabetes and infections was found (all of which are amongst the 10 leading causes of death, WHO, 2017–Fig.1). Total mortality was reduced considerably by up to 16% for both men and women who drank 4–5 cups of coffee a day. Similar associations were observed whether participants drank predominantly caffeinated or decaffeinated coffee (Folmer et al., 2017). Although scientific studies can link certain compounds to specific mechanisms, it is likely that most contributions to decreasing the risk of certain diseases are caused by synergistic or additive effects with various compounds present in coffee. The next section presents summaries of research results on the effect of coffee and health, exploring the most studied effects individually.

3.1 Coffee consumption for mental and physical performance and well-being

The stimulating effect of coffee is well known and is due to caffeine's ability to enhance mental performance, which includes enhancing alertness and perception (Einother and Giesbrecht, 2012). According to the EFSA, who reviewed existing evidence of caffeine on mental performance (EFSA 2045, 2011c), generally, a dose of 75 mg is required to obtain these effects, although very large differences in individual responses to caffeine are observed. Caffeine consumption can also improve other functions such as memory (Nehlig, 2010; Borota et al., 2014) and mood (Smith, 2005; Olson et al., 2010). Coffee components other than caffeine have also been shown to influence cognitive performance in an elderly population, though to a smaller extent than caffeine. Decaffeinated coffee enriched in chlorogenic acids can improve alertness and reduce headaches and mental

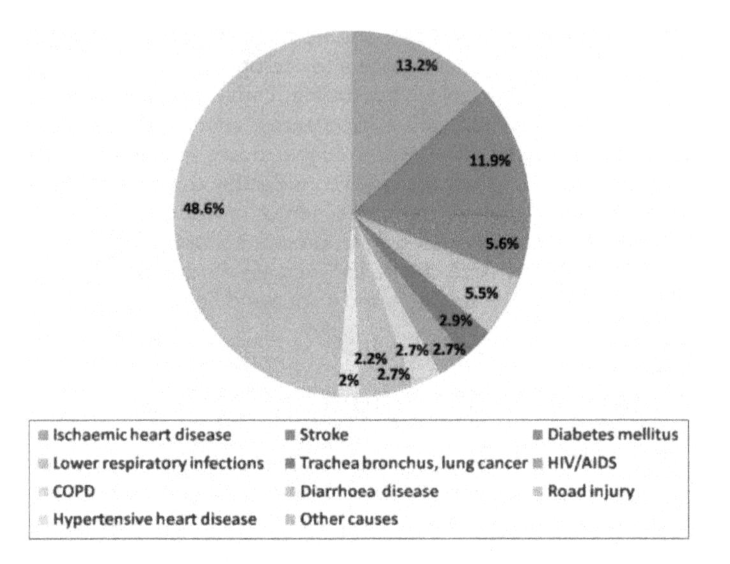

Figure 1 The ten leading causes of death in the world by percentage (data from WHO, 2017, updated in 2014).

fatigue in comparison to non-enriched decaffeinated coffee. The effects may be partly attributed to chlorogenic acids, but other compounds naturally present in coffee are also suggested to play a role (Camfield et al., 2013; Cropley et al., 2012; Folmer et al., 2017).

The large inter-individual variability of the stimulating effects of caffeine is due to the difference in the ability to metabolize and eliminate it from the body. Whilst for most people, it takes about three to six hours to eliminate 50–75% of the caffeine and its metabolites (Goldstein, 2010), for some people, it can take much longer. The effects of several cups of coffee on these individuals, usually called 'slow metabolizers', may therefore be accumulative for a while. The variability in the enzymatic breakdown of caffeine may account for its variable effect on sleep induction and arousal (Youngberg et al., 2011). The stimulating effects of caffeine tend to be stronger when the individual is in a state of fatigue or in elderly people (van Boxtel and Schmitt, 2004). However, habitual caffeine consumers may suffer less from these issues as they develop a tolerance. In this case, caffeine will, for example, still disrupt their sleep, but to a lesser extent than for people who are not habitual consumers (Childs and de Wit, 2012; Drapeau, 2006). It is the responsibility of each person to pay attention to his or her response to caffeine intake at different times of the day, and adapt intake patterns accordingly.

Coffee and other caffeine vehicles have been used by athletes for a long time, and the initial papers discussing the mechanisms involved date back to 1978 (Costill, 1978). More specifically, caffeine exerts a positive effect on the endurance and exercise capacity, due to the effect on neural mechanisms (Spriet and Gibala, 2004;Folmer et al., 2017). Caffeine also seems to reduce the pain perception due to an increase in the secretion of β-endorphins which exhibit analgesic properties (O'Connor et al., 2004). It is well documented that caffeine can enhance endurance and coordination, stop–go events (e.g. team and racket sports) and sports involving sustained high-intensity activity lasting from

1 min up to an hour (e.g. swimming, rowing and running races) (Jenkins, 2008; Hogervost et al., 2008; Folmer et al., 2017).

Based on the scientific studies, the active dose of caffeine was found to be 3 mg per kg of bodyweight, to be taken 1 h before exercise (EFSA, 2054, 2011c; Goldstein et al., 2010). For a person weighing about 70 kg, this amount would thus be equivalent to 210 mg. According to the EFSA caffeine safety report (EFSA 4102, 2015a), it is safe to consume single doses of 200 mg of caffeine less than 2 h prior to intense exercise. However, the amount and time prior to exercise for an optimal effect will vary for different individuals due to differences in metabolic rates (Folmer et al., 2017). In 1994, caffeine was removed from the list of banned substances.

The impact of caffeine on the mental and physical health of women and children is a more recent area of interest, even if the initial papers appeared in the early 1990s. Evidence suggests that the normal hormonal changes during pregnancy slow the body's ability to metabolize caffeine. Therefore, a given dose of caffeine can have longer-lasting effects (as long as 15 h in the third trimester) (Kuczkowski, 2009). Even though the EFSA report on caffeine safety (EFSA 4102, 2015a) concludes that its consumption is safe for pregnant and lactating women, it recommends an intake reduction to a maximum of 200 mg throughout the day. Based on scientific findings, there is no risk of adverse birth weights for caffeine consumption below these values. Nevertheless, the risks of very high intake (more than 600 mg of caffeine per day) include foetal growth retardation and low weight for gestational age (Sengpiel et al., 2013). Although there is no consensus in studies suggesting that caffeine could delay the time of conception, it may be prudent for women who have difficulty in conceiving to limit the caffeine intake to less than 300 mg per day (Higdon and Frei, 2006; Folmer et al., 2017).

It is known that caffeine is present in the milk of lactating coffee drinkers with a peak appearing about 1 h after consuming a caffeinated beverage (Stavchansky et al., 1988; Nehlig and Debry, 1994). For this reason, doctors recommend that breastfeeding women keep caffeine intake below 200 mg per day (EFSA 4102, 2015a). At these levels, studies show that the sleep time of nursing infants is similar to controls (Santos et al., 2012; Clarc and Landholt, 2016; Folmer et al., 2017).

When it comes to children and coffee consumption, there are major cultural differences in both overall coffee consumption and consumption guidelines. For example, in most European countries, habitual coffee consumption starts when children become adults, and until the age of 10, chocolate and tea are the main sources of caffeine (EFSA 4102, 2015a). Brazil has implemented an active coffee school programme based on the findings that 20% coffee added to a glass of whole milk helps children perform better in school (ABIC, 2016). Additionally, there are studies that show that caffeine may attenuate the symptoms of attention-deficit syndrome (Garfinkel et al., 1981). European adolescents consume less coffee and their source of caffeine intake is widely distributed amongst different types of food and beverages (EFSA 4102, 2015a). In the United States, this section of the population primarily consumes caffeine from soft drinks (USDA, 2015).

As information on the impact of caffeine on the health of children and adolescents is scarce, it is difficult to derive general conclusions on safe intake levels. Caffeine doses of about 1.4 mg per kg bodyweight or more may impact sleep quality in adults, particularly when consumed close to bedtime (EFSA 4102, 2015a). For this reason, and because data on safe habitual caffeine intake for children and adolescents are insufficient, the EFSA suggests a limit of 3 mg of caffeine per kg of bodyweight per day (EFSA 4102, 2015a), which would equal around 90 mg for a 10-year old (Folmer et al., 2017).

Canadian authorities are more conservative and suggest a limit of 2.5 mg per kg of bodyweight per day (Health Canada, 2011). The short-term risk associated with children and caffeine consumption is that caffeine may cause anxiety and nervousness (Nawrot et al., 2003).

3.2 Coffee and cognitive health

The acute effects of caffeine were discussed in the mental performance section. In this section, we will look at the long-term effects of coffee on reducing the risk of cognitive degenerative diseases. Cognitive functions such as verbal ability, inductive reasoning and perceptual speed decrease after 20 years of age. Genetics, life events and lifestyle factors impact the rate and amplitude of this decline (Hedden and Gabrieli, 2004; Folmer et al., 2017). A large number of epidemiological studies relate the regular consumption of coffee to a reduced appearance of cognitive decline in the elderly (Arab et al., 2013; Ritchie, 2007; Corley, 2010). A meta-analysis of these human studies suggests that there is a clear protective effect of caffeine consumption, rather than from coffee itself (Santos et al., 2010; Ryan, 2002).

Alzheimer's disease is the most frequent cause of dementia, leading to a progressive cognitive decline. Whilst there is currently no medication for Alzheimer's disease (Waite, 2015), there are studies that show an inverse association between the coffee consumption and the development of Alzheimer's disease, with a 27% risk reduction (Barranco Quintana et al., 2007; Waite, 2015; Folmer et al., 2017). The mechanism is believed to be related to the anti-inflammatory effect of caffeine on the A1 and A2 receptors, in addition to reducing the deposits of toxic beta-amyloid peptide in the brain, a pathological characteristic in patients with Alzheimer's disease (Rosso, 2008; Arendash and Cao, 2010). In addition to caffeine, the intake of polyphenols also seems to help decrease the risk of Alzheimer's disease. Emerging evidence from animal models also links chlorogenic acids to the prevention of neurodegenerative disease and ageing (Esposito et al., 2002; Ramassamy, 2006). Although the involvement of coffee polyphenols in the human cognitive function has not been well studied, the number of findings on the *in vitro* neuroprotective effects of polyphenols in general is rapidly increasing (Lakey-Beitia, et al., 2015). Initial indications relate the anti-inflammatory effects of polyphenols to the reduced risk of developing Alzheimer's disease. Other proposed mechanisms could be i) inhibition of the enzymes acetylcholinesterase and butyrylcholinesterase in the brain, as this retards acetylcholine and butyrylcholine breakdown and ii) the prevention of oxidative stress–induced neurodegeneration due to its high antioxidative activity (Oboh et al., 2013, Folmer et al., 2017).

Similar to Alzheimer's disease, a large number of epidemiological studies have reported an inverse relationship between the caffeine consumption and the risk of developing Parkinson's disease. The latter is a neuropathological disorder that slows down the motor function, whilst generating resting tremors, muscular rigidity, gait disturbances and impairing postural reflex. It involves the degeneration of neurons in the brainstem (Kuwana et al., 1999). Coffee consumption appears to reduce, or delay, the development of Parkinson's disease. From the meta-analysis of 26 studies, a 25% lower risk of Parkinson's disease was found in coffee drinkers compared with non-coffee drinkers. The mechanism is probably related to the capacity of caffeine to block the A2 adenosine receptors in the brain (Costa et al., 2010). Studies recently outlined a possible additional mechanism. A rodent model showed that trigonelline may exert a neuroprotective effect, inducing a significant reversal of motor dysfunction (Nathan et al., 2014; Folmer et al., 2017).

3.3 Coffee and cardiovascular disease

One of the misinterpretations linking coffee and health stemmed from the belief that the risk of cardiovascular disease, the leading cause of death in the world according to the WHO (13.2% and 2% of deaths due to ischaemic and hypertensive cardiovascular diseases, respectively, WHO, 2017;Fig.1), was increased by drinking coffee. This belief was supported by the fact that caffeine increases blood pressure and acutely reduces insulin sensitivity after coffee consumption. However, it is now known that most acute caffeine effects cease to exist with regular coffee consumption due to adaptation mechanisms and that other coffee components, mainly chlorogenic acids and trigonelline, have compensatory effects on endothelial dysfunction and insulin resistance. Additionally, *in vitro* and animal studies indicate that coffee has high antioxidant and anti-inflammatory potential, improves endothelial dysfunction and reduces insulin resistance, which are key mechanisms for cardiovascular protection (Rebello and Van Dam, 2013).

Corroborating these findings, dozens of studies have shown the inverse association between coffee consumption and cardiovascular diseases. Andersen et al. (2006) studied the relationship of coffee drinking with total mortality and mortality attributed to cardiovascular disease, cancer and other diseases with a major inflammatory component. A total of 41 836 postmenopausal women aged 55–69 years at baseline were followed for 15 years. During this period, there were 4265 deaths. Evaluating the causes of mortality, the authors observed that coffee consumption increasingly reduced the risk of cardiovascular and other inflammatory diseases in postmenopausal women, thereby decreasing mortality from these diseases. This effect was attributed to the ability of coffee to inhibit inflammatory processes via its antioxidative and anti-inflammatory compounds.

A meta-analysis was carried out by Crippa et al. (2014), using 21 prospective studies, with 997 464 participants and 121 915 reported deaths. Results indicated that coffee consumption is inversely associated with all-cause and cardiovascular disease mortality and that the risk was increasingly reduced for those who consumed 3 to 4 cups. Similar results were observed by Malerba et al. (2013).

Another recent study by Ding et al. (2015) examined the causes of death of 19 524 women and 12 432 men from two large cohort studies in the United States, the Harvard Health Professionals Follow-up Study and the Nurses' Health Study (1 and 2). Inverse associations were observed between the consumption of regular and decaffeinated coffee and the deaths due to cardiovascular and neurological diseases. When restricting to those that had never smoked, the all-cause mortality risk was also increasingly reduced as the number of cups increased; however, higher consumption reduced the benefit somewhat.

Stroke is the second leading cause of death in the world as estimated by the WHO (11.9% of the total deaths, WHO, 2017; Fig.1); however, data on the association between coffee consumption and risk of stroke are scarce. A study by Lopez-Garcia (2009) analysed data from a prospective cohort of 83 076 women in the Nurses' Health Study for 24 years. Results evidenced that long-term coffee consumption is not associated with an increased risk of stroke in women. In contrast, data suggested that coffee consumption may modestly reduce this risk. Decaffeinated coffee was associated with a trend towards a lower risk of stroke after adjustment for caffeinated coffee consumption. Using data from 11 prospective studies with 479 689 participants and 10 003 cases of stroke, a meta-analysis performed by Larsson and Orsini (2011) corroborated the results obtained for women, finding an inverse, although modest, association between moderate coffee consumption and risk of stroke.

3.4 Coffee and type2 diabetes

Diabetes mellitus is characterized by a high blood glucose level, which can cause complications such as cardiovascular diseases, stroke, chronic kidney failure, foot ulcers and damage to the eyes (IDF, 2012). There are three main types of diabetes: type 1, in which the pancreas fails to produce enough insulin and which is generally genetically determined; type 2 diabetes is the seventh leading cause of death in the world (2.7% of the universal deaths, WHO, 2017), starts with insulin resistance (lack of insulin may also develop) and is promoted by obesity and a sedentary lifestyle (Coope et al., 2015); and gestational diabetes, an often transient disease that occurs when pregnant women develop a high blood sugar level (IDF, 2015; Folmer et al., 2017).

Floegel et al. (2012) investigated the association between the coffee consumption and the risk of chronic diseases, including type 2 diabetes. They used data from 42 659 participants collected over 8.9 years from the European Prospective Investigation into Cancer and Nutrition (EPIC) cohort. They found an inverse association of consumption of more than 4 cups (150 mL) of regular coffee per day with the overall risk of type 2 diabetes.

A number of similar studies have observed such effects related to regular coffee drinking. A recent meta-analysis of large epidemiological studies confirmed the link between moderate coffee consumption and a reduced risk of developing type 2 diabetes across different populations (Ding et al., 2014). The findings from these systematic studies demonstrate a clear inverse association between the coffee consumption and the risk of developing diabetes. Compared with no, or infrequent, coffee consumption, the risk of developing type 2 diabetes was reduced linearly, with a 33% reduction for 6 cups per day. In a similar comparison, drinking up to four cups per day of decaffeinated coffee was associated with a 20% reduced risk (Ding et al., 2014). This suggests that the protective effects of coffee on diabetes are independent of caffeine.

Animal studies have indicated that the main compounds responsible for the protective effect are chlorogenic acids (Kempf, 2010) and its derivatives, as well as trigonelline (van Dijk et al., 2009; Rios et al., 2015). They appear to preferentially target hepatic glucose metabolism by improving insulin sensitivity (Lecoultre et al., 2014). Other proposed mechanisms observed in *in vivo* and *in vitro* studies include the regulation of key enzymes of glucose and lipid metabolism, such as glucokinase, glucose-6-phosphatase, fatty acid synthase and carnitinepalmitoyltransferase (Waite, 2015). In a human study, trigonelline generated significantly lower glucose and insulin levels after an oral glucose load compared with a placebo (Rios et al., 2015).

3.5 Coffee and liver diseases

There are a number of diseases that can impact liver health and include both liver cancer and cirrhosis, a progressive disease caused by liver steatosis (fatty liver) and alcohol abuse, where the healthy tissue is replaced by the scar tissue and eventually prevents the liver from functioning correctly (Saab et al., 2014). According to a recent meta-analysis of 16 human studies, coffee consumption reduces the risk of developing liver cancer by 40% compared with no coffee consumption (Larsson and Wolk, 2007; Bravi et al., 2013).

In a clinical study performed in Brazil, caffeine consumption greater than 123 mg per day was also associated with reduced hepatic fibrosis (Machado et al., 2014). In addition, the study observed positive effects of regular coffee consumption in patients with chronic hepatitis C.

A number of *in vitro* studies have demonstrated the strong role of the chlorogenic acids present in coffee in protecting the liver from damage at various levels, possibly by preventing cell apoptosis and oxidative stress damage due to the activation of the body's natural antioxidant and anti-inflammatory systems (Ji et al., 2013). Coffee melanoidins have also been reported to have a protective effect on liver steatosis in obese rats (Vitaglione et al., 2012), which suggests that the melanoidins in coffee may have an influence on liver fat and functionality. Although melanoidins do not seem to be absorbed in humans, they can function as an antioxidant dietary fibre, like the unabsorbed portion of chlorogenic acids, quenching radicals and improving the reduced/oxidized glutathione balance in the colon. At the same time, they may act to promote the growth of a beneficial colon microbiota, affecting inflammatory pathways in the colon and consequently in the liver (Folmer et al., 2017).

3.6 Coffee and cancer

In the broadest sense, cancer represents the final result of abnormal cell growth and can occur in most human tissues. The carcinogenicity of coffee drinking was assessed by the IARC in 1991. At that time, coffee was classified as 'possibly carcinogenic to humans' (Group 2B), based on limited evidence of an association with cancer of the urinary bladder from case-control studies, and inadequate evidence of carcinogenicity in experimental animals. When subsequent studies were controlled for smoking, they failed to show an elevated risk of bladder cancer (Butt and Sultan, 2011). Recently, the IARC re-evaluated the carcinogenicity of drinking coffee and other hot beverages (Loomis et al., 2016), using a much larger database of more than 1000 observational and experimental studies. In assessing the accumulated epidemiological evidence, more weight was given to well-conducted prospective cohort and population-based case-control studies that controlled adequately for important potential confounding factors, including smoking (tobacco) and alcohol consumption. In conclusion, there was no consistent evidence associating drinking coffee with bladder cancer. In contrast, for endometrial cancer, the five largest cohort studies showed mostly inverse associations with coffee drinking. These results were supported by the findings of several case-control studies and a meta-analysis. Inverse associations with coffee drinking were also observed in cohort and case-control studies of liver cancer in Asia, Europe and North America. A meta-analysis of prospective cohort studies estimated that the risk of liver cancer decreases proportionally with coffee intake. No association or a modest inverse association for female breast cancers was found. Similarly, no association was found for pancreas and prostate cancers. Data were also available for more than 20 other cancers, including lung, colorectal, stomach, oesophageal, oral cavity, ovarian and brain cancers and childhood leukaemia. Although the volume of data for some of these cancers was substantial, evidence was inadequate for all the other cancers reviewed for reasons including inconsistency of findings across studies, inadequate control for potential confounding factors, potential for measurement error, selection bias or recall bias or insufficient numbers of studies (Loomis et al., 2016). As a result of this re-evaluation, coffee was upgraded by the IARC and is no longer considered to be potentially carcinogenic. In summary, epidemiological data demonstrated that coffee consumption is actually associated with a *lower* overall risk of cancer, especially liver and endometrial cancers.

There are several compounds in coffee that have been found to play a protective role against cancer, and the most well-known are chlorogenic acids and their derivatives. The contribution of melanoidins has also been suggested to decrease the risk of colon cancer (see chlorogenic acids and melanoidins topics in this chapter). Based on *in vitro* and animal evidence, coffee diterpenes are also strong candidates.

Epidemiologic studies have shown an increased risk of oesophageal cancer from drinking hot beverages such as maté, tea or coffee. It has been observed that the intra-oesophageal temperature is increased by 6–12°C when coffee was drunk at 65°C (Islami et al., 2009). The high-temperature injures the oesophageal mucosa and consequently causes inflammation or forms reactive nitrogen species, a type of free radical. It has been suggested by the IARC that drinking coffee, and other hot beverages, at temperatures above 65°C increases the risk of oesophageal cancer (Loomis et al., 2016). Although in some countries, coffee is consumed at temperatures below 65°C, in other countries, the temperature can be much higher. In public places, serving coffee at very high temperatures may influence people to drink it hotter than they would at home, so people should be aware of this factor for all hot beverages, soups and hot foods in general.

4 Potential side effects of coffee drinking

4.1 Hyper stimulation and sleep quality and duration by caffeine

Caffeinated coffee can cause irritability and anxiety, and reduce sleep quality by increasing the time required to fall asleep, interfering with the depth of sleep and reducing the total time spent sleeping. It can also cause more frequent awakening or sleep fragmentation (Folmer et al., 2017; Clarc and Landholt, 2016; Huang et al., 2011). The use of caffeine in energy drinks, and the risk of overdosing in children, motivated health authorities to evaluate and publish guidelines on safe caffeine consumption. The most recent report is the EFSA's 2015 scientific opinion on caffeine safety (EFSA 4102, 2015a). National health authorities have also published reports like the US Department of Agriculture report (2015). The general agreement is that the habitual consumption of up to 400 mg of caffeine per day, and up to 200 mg per serving, does not cause safety concerns for non-pregnant adults. Considering a range between 100 and 200 mg caffeine per cup, this would translate into 2–4 cups per day.

4.2 Caffeine tolerance, dependence and withdrawal

Caffeine is the most widely used psychoactive substance in the world, and the issue of possible dependence on caffeine has been discussed for many years. In fact, different drugs affect different people in different ways, and caffeine is no exception. It is therefore difficult to make general statements on dependence, tolerance and withdrawal; however, there is no such brain circuit that links caffeine to dependence. Caffeine does not affect areas involved in reinforcing and rewarding (Nehlig, 2010). According to the standard for measuring any potential drug abuse and dependence (as defined by the Diagnostic and Statistical Manual of Mental Disorders (DSM-IV, American Psychiatric Association, 2000), there are no criteria that qualify caffeine for potential drug abuse (Folmer et al., 2017).

As with any drug, regular caffeine users will establish a partial tolerance to caffeine. However, studies have shown that this tolerance only applies to effects such as jitteriness, anxiety and an increased heart rate. Users do not develop a tolerance to the benefits of caffeine consumption such as improved mental performance, although sometimes slightly higher doses of caffeine are required (Satel, 2006).

The types of caffeine withdrawal symptoms which are most often reported are headaches; feelings of weariness, weakness and drowsiness; impaired concentration; fatigue and work difficulty; depression; anxiety; irritability; increased muscle tension and occasional tremors, nausea or vomiting. Withdrawal symptoms generally peak 20–48 h after the last caffeine was consumed, although users can generally avoid these if caffeine consumption is progressively decreased (Nehlig, 2010; Folmer et al., 2017).

Excessive coffee intake does not cause organic toxicity, but it can generate negative side effects, such as those associated with caffeine withdrawal. Symptoms related to the toxicity of coffee can occur at levels well below fatal doses; for example, concentrations above 15 mg caffeine per kg of bodyweight may be toxic for the cardiovascular, nervous and gastrointestinal systems (e.g. 1 g of caffeine for a person weighing 70 kg). Although such caffeine levels are not easily obtained through acute coffee intake, users may easily consume caffeine pills in such quantities. Reported overdose symptoms are hypertension or hypotension, tachycardia, vomiting, fever, delusion, hallucinations, arrhythmia, cardiac arrest, coma and death. Fatalities most commonly result from seizures and cardiac arrhythmias at plasma levels of 100–180 μg per mL (Childs and de Wit, 2012; Folmer et al., 2017). However, caffeine-related deaths have not been associated with coffee drinking (Yamamoto, 2015).

4.3 Cholesterol-raising effects of diterpenes

Epidemiologic and mechanistic studies have reported that the diterpenes, cafestol, and to a lesser extent, kahweol, naturally found in coffee oil and in unfiltered coffees, can alter lipid enzymes and thus influence cholesterol levels. This relationship was found to be linear with increasing cafestol consumption (Urgert and Katan, 1997). A meta-analysis of a set of 18 clinical intervention trials on coffee consumption and cholesterol and serum lipids was performed by Jee et al. (2001). The authors corroborated the dose–response relationship between coffee consumption and cholesterol and observed a strong increase upon the consumption of 6 or more cups of boiled coffee per day, which was not observed when a paper filter was used.

The high consumption of diterpenes has been associated with elevated homocysteine and low-density lipoprotein levels in human plasma, which may indirectly increase the risk of cardiovascular diseases (Farah, 2012).

4.4 Gastro-oesophageal reflux or heartburn

Gastro-oesophageal reflux, also called heartburn, is caused by the reflux of gastric fluid into the oesophagus due to the low pressure in the sphincter muscle at the junction of the stomach and oesophagus. A number of people suffering from this condition have mentioned that coffee may be one of the food products causing this complaint. Although some studies have tried to investigate this, the role of coffee consumption in reflux is still unclear. It has, for example, been reported that a few compounds in coffee stimulate the production of gastric juice, which is very acidic when first produced in the stomach (pH 1–2). Chlorogenic acids, Nβ-alkanoil-5-hydroxytryptamides (C5HTs) from coffee wax, and to a lesser extent, caffeine, are some of the main compounds thought to promote this effect (Fehlau and Netter, 1990). Additionally, it has been hypothesized that the roasting products of chlorogenic acid such as pyrogallol, like C5HTs, irritate the gastric mucosa (Darboven, 1997). In addition to the stimulation of gastric juice, coffee consumption also seems to cause muscle contraction impairment (relaxation effect) of the lower oesophageal sphincter in some individuals, promoting heartburn, which has been attributed to caffeine (Terry et al., 2000). The pH of a coffee brew is mildly acidic, commonly fluctuating between 5.8 and 5.5 in robusta coffees and between 4.3 and 4.8 in fresh lightly roasted acidic arabica coffees, with approximately pH 5.0 being more usual in dark roasted blends. This is much higher than the pH of gastric juice or, for example, the pH of apple juice (pH 4.3–3.3) or citric juice (pH 2.3–3.3) (Farah, unpublished).

Therefore, based on current knowledge, the stimulation of gastric juice production along with the relaxation of the oesophageal sphincter seems to be the most likely causes for heartburn, although some studies support the contribution of the acidity of foods to heartburn (Feldman and Barnett, 1995). Since there are only a few studies on this subject and none in humans, more mechanistic and clinical studies are necessary to prove the involvement of each of these specific coffee compounds in this disease, as well as improving conditions in their absence.

Some pre-roasting technological methods have been developed which aim to decrease heartburn, although no clinical studies have yet proved that these treatments are effective in humans. Reducing coffee wax, and thus C5HTs, can be achieved by applying steam treatment, either as a stand-alone process, or as part of the water or CO_2 decaffeination methodologies (which in addition decreases the content of chlorogenic acids and caffeine).

5 Final considerations

Since the initial studies published in medical journals in the eighteenth century, coffee has been through many waves of approval and disapproval. As science has evolved and confounding factors could be accounted for, an increasing number of studies have found correlations between coffee consumption and reduced risk of developing certain diseases. As *in vitro* and animal studies confirm the involvement of active coffee components in specific diseases, the challenge remains to fully understand the mechanisms that these active compounds exert, as coffee is a molecularly highly complex beverage made up of thousands of compounds. It is, however, becoming more evident that it is not only the specific compounds in coffee, but rather the beverage as a whole that is responsible for its beneficial effects.

Despite the potentially positive contribution of coffee to reducing the risk of certain diseases, these findings need to be related to each other and, more importantly, to lifestyle factors that influence the risk of developing certain diseases and longevity. Some of these include not smoking, good nutrition (a balanced and varied diet including five servings of fruit and vegetables daily), exercise, low alcohol consumption and low stress, all of which have a strong documented impact on disease prevention and life expectation (Khaw et al., 2008).

6 Acknowledgements

The author would like to thank Britta Folmer, from Nestlé Nespresso SA, for her valuable contribution to this chapter. The Research scholarships provided by the Brazilian National

Council for Scientific and Technological Development – CNPq and the Research Support Foundation of Rio de Janeiro – FAPERJ are greatly appreciated.

7 Where to look for further information

7.1 Recommended books

Folmer, B. (Ed). (2017). *The Craft and Science of Coffee*. Elsevier, London,1st edition.
Farah, A. (Ed). (2017). *Coffee: Chemistry Quality and Health Implications*. Royal Society of Chemistry, UK, 1st edition, In press.
Preedy, V. (Ed). (2015). *Coffee in Health and Disease Prevention*. Elsevier, London, UK, 1st edition.
Feng, I. (2012). *Coffee: Emerging Health Effects and Disease Prevention*. IFT Press/Willey-Blackwell, USA.

7.2 Recommended websites

- International coffee organization (ICO): www.ico.org. The International Coffee Organization (ICO) is the main intergovernmental organization for coffee, bringing together producing and consuming countries to tackle the challenges faced by the world coffee sector through international cooperation. It makes a practical contribution to the world coffee economy and to improving the standards of living in developing countries by enabling government representatives to exchange views and coordinate coffee policies and priorities, and enabling government representatives to exchange views and coordinate coffee policies and priorities at regular high level. On this site, in addition to the information on world coffee production, exports and imports, general global information on coffee is also found.
- Association for Science and Information on Coffee (ASIC): www.asic-cafe.org. 'ASIC is a completely independent organization in the world whose scientific vocation is specifically devoted to the coffee tree, the coffee bean and the coffee drink'. On ASIC's site, you will find the proceedings of the previous colloquia as well as information and links on the latest publications on coffee agronomy, chemistry, technology, coffee and health and physiological effects of coffee.
- The Specialty Coffee Association (SCA) that is a membership-based association acts as a unifying force within the specialty coffee industry and works to make coffee better by raising standards worldwide through a collaborative and progressive approach. Members of the SCA include coffee retailers, roasters, producers, exporters and importers, as well as manufacturers of coffee equipment and related products. For more information, access www.sca.coffee.The SCA was formed in January 2017 following the merger of the SCAA and SCAE, acting in the United States and Europe. The respective websites are currently still active and information on training, education, events and standards can still be found (seewww.scaa.org and www.scae.org)
- A website fully dedicated to entire information on coffee and health is www.coffeeandhealth.org. It is a science-based resource developed for health care and other professional audiences and provides the latest information and research into coffee, caffeine and health.

8 References

ABIC, 2016 http ://abic.com.br/institucional/projetos-sociais/ accessed dez, 2016 and 2017.

Alcázar, A., Fernández-Cáceres, P. L., Martın-Valero, M., Pablos, F. and González, A. G., 2003. Ion chromatographic determination of some organic acids, chloride and phosphate in coffee and tea. *Talanta, Amsterdam* 61(1), 95–101.

American Psychiatric Association. 2000. *Diagnostic and Statistical Manual of Mental Disorders*, 4th Ed. American Psychiatric Association, Arlington, VA.

Antonio, A. G., Moraes, R. S., Perrone, D., Maia, L. C., Santos, K. R. N., Iório, N. L. P. and Farah, A., 2010. Species, roasting degree and decaffeination influence the antibacterial activity of coffee against Streptococcus mutans. *Food Chemistry* 118, 782–8.

Andersen, L. F., Jacobs Jr., D. R., Carlsen, M. H. and Blomhoff, R., 2006. Consumption of coffee is associated with reduced risk of death attributed to inflammatory and cardiovascular diseases in the Iowa Women's Health Study. *American Journal of ClinicalNutrition* 83(5), 1039–46.

Arab, L., Khan, F' and Lam, H., 2013. Epidemiologic evidence of a relationship between tea, coffee, or caffeine consumption and cognitive decline. *Advances in Nutrition* 4(1), 115–22.

Arauz, J., Rivera-Espinoza, Y., Shibayama, M. and Muriel, P., 2015. Nicotinic acid prevents experimental liver fibrosis by attenuating the prooxidant process. *International Immunopharmacology* 28(1), 244–51.

Arendash, G. W. and Cao, C., 2010. Caffeine and coffee as therapeutics against Alzheimer's disease. *Journal of Alzheimer's Disease* 20(S1), 117–26.

Balzer, H. H., 2001. Acids in coffee. In: *Coffee Recent Developments* (Clarke, R. J. and Vitzthum, O. G. (Eds). Blackwell Science, Berlin, p. 18.

Barranco Quintana, J. L., Allam, M. F., SerranoDel Castillo, A. and Fernandez-Crehuet Navajas, R., 2007. Alzheimer's disease and coffee: A quantitative review. *Neurological Research* 29, 91–5.

Bizzo, M. L. G., Farah, A., Kemp, J. A. and Scancetti, L. B., 2015. Highlight in the history of coffee science related to health. In: *Coffee in Health and Disease Prevention* (Preedy, V. (Ed.)). Elsevier, London, UK, pp.11–18.

Borrelli, R. C., Esposito, F., Napolitano, A., Ritieni, A. and Fogliano, V., 2004. Characterization of a new potential functional ingredient: Coffee silverskin. *Journal of Agriculture and Food Chemistry* 52, 1338–43.

Boekshoten, M.V., Van Cruschten, S.T., Kosmeijer-Schuil, T.G. and Katan, M.B. 2006. Negligible amounts of cholesterol-raising diterpenes in coffee made with coffee pads in comparison with unfiltered coffee, 2006. *Nederlands Tijdschrift voor Geneeskunde*, 150, 2873–5.

Borota, D., Murray, E., Kecell, G., Chang, A., Watabe, J. M., Ly, M., Toscano, J. P. and Yassa, M.A., 2014. Post-study caffeine administration enhances memory consolidation in humans. *Nature Neuroscience* 17, 201–3.

Bravi, F., Bosetti, C., Tavoni, A., Gallus, S. and La Vecchia, C., 2013.Coffee reduces risk for hepatocellular carcinoma: An updated meta-analysis. *Clinical Gastroenterology and Hepatology* 11, 1413–21.

Butt, M. S. and Sultan, M. T., 2011. Coffee and it's Consumption: Benefits and Risks. *Critical Reviews in Food Science and Nutrition* 51, 363–73.

Camfield, D. A., Silber, B. Y., Scholey, A. B., Nolidin, K., Goh, A. and Stough, C., 2013. A Randomised placebo-controlled trial to differentiate the acute cognitive and mood effects of chlorogenic acid from decaffeinated coffee. *Public Library of Science One* 8(12), e82897.

Carpenter, K. J., 1983. The relationship of pellagra to corn and the low availability of niacin in cereals. *Experientia Suppl.* 44: 197–222.

Casal S., 2017. Potential effects of β-carbolines on human health. In *Coffee: Chemistry, Quality, and Health Implications* (Farah, A. (Ed.)). Royal Society of Chemistry, London, UK. In press.

Chang, W. H., Hu, S. P., Huang, Y. F., Yeh, T. S. and Liu, J. F., 2010. Effect of purple sweet potato leaves consumption on exercise-induced oxidative stress and IL-6 and HSP72 levels. *Journal of Applied Physiology* 109(6), 1710–15.

Childs, E. and de Wit, H., 2012. Potential mental risks. In: *Coffee, Emerging Health Effects and Disease Prevention* (Chu, Y.-F. (Ed.)). Wiley-Blackwell, Oxford, UK, pp. 293–306.

Clarc, I. and Landholt, H. P., 2017. Coffee, caffeine and sleep: A systematic review of epidemiological studies and randomized controlled trials. *Sleep Medicine Reviews* 31, 70–8.

Cliford, M. N. 1985. Chlorogenic acids in coffee. In *Chemistry*, vol 1, Clarke R. J. and Macrae R. (Eds), Elsevier Applied Science, London, 153–202.

Coope, A., Torsoni, A. S. and Velloso, L., 2015. Mechanisms in endocrinology: Metabolic and inflammatory pathways on the pathogenesis of type 2 diabetes. *European Journal of Endocrinology* 15, 1065.

Corley, J., Jia, X., Kyle, J. A., Gow, A. J., Brett, C. E., Starr, J. M., McNeill, G. and Deary, I. J. 2010. Caffeine consumption and cognitive function at age 70: The Lothian Birth Cohort 1936 study. *Psychosomatic Medicine* 72, 206–14.

Costa, J., Lunet, N., Santos, C., Santos, J. and Vaz-Cameiro, A., 2010. Caffeine exposure and the risk of Parkinson's disease: A systemic review and meta-analysis of observational studies. *Journal of Alzheimer's Disease* 20, S221–38.

Costill, D. L., Dalsky, G. P. and Fink, W. J., 1978. Effects of caffeine ingestion on metabolism and exercise performance. Medicine *Science in Sports* 10(3), 155–8.

Crippa, A., Discacciati, A., Larsson, S. C., Wolk, A. and Orsini, N., 2014. Coffee consumption and mortality from all causes, cardiovascular disease, and cancer: A dose-response meta-analysis. *American Journal of Epidemiology* 180, 763–75.

Cropley, V., Croft, R., Silber, B., Neale, C., Scholey, A., Stough, C. and Schmitt, J., 2012. Does coffee enriched with chlorogenic acids improve mood and cognition after acute administration in healthy elderly? A pilot study. *Psychopharmacology* 219(3), 737–49.

Cunha, S. and Fernandes, J. 2017. Pesticides. In: *Coffee: Chemistry, Quality and Health Implications* (Ed.: Farah, A.). Royal Society of Chemistry, London, UK. In press.

Darboven, A.1997. Method for the quality improvement of raw coffee by treatment with steam and water (in German). Europaisches Patent- € blatt, 1997/05/95109295.6.

Ding, M., Satija, A., Bhupathiraju, S. N., Hu, Y., Sun, Q., Han, J., Lopez-Garcia, E., Willet, W., van Dam. R. M. and Hu, F. A., 2015. Association of coffee consumption with total and cause-specific mortality in three large prospective cohorts.*Circulation* 132(24), 2305–15.

Ding, M., Bhupathiraju, S.N., Chen, M., vanDam, R. M. and Hu, F. B., 2014. Caffeinated and decaffeinated coffee consumption and risk of Type 2 diabetes: A systematic review and a dose-response meta-analysis. *Diabetes Care* 37(2), 569–86.

Drapeau, C., Hamel-Hebert, I., Robillard, R., Seimaoui, B., Filipini, D. and Carner, J., 2006. Challenging sleep in aging: the effects of 200 mg of caffeine during the evening in young and middle-aged moderate caffeine consumers. *Journal of Sleep Research* 15(2), 133–41.

Einother, S. J. L. and Giesbrecht, T., 2012. Caffeine as an attention enhancer: Reviewing existing assumptions. *Psychopharmacology* 225(2), 251–74.

Esposito, E., Rotilio, D., Di Matteo, V., Di Giulio, C., Cacchio, M. and Algeri, S., 2002. A review of specific dietary antioxidants and the effects on biochemical mechanisms related to neurodegenerative processes. *Neurobiology of Aging* 23, 719–35.

EFSA (European Food Safety Authority), 2008. Scientific opinion of the panel on contaminants in the food chain on a request from the European commission on polycyclic aromatic hydrocarbons in food. *The EFSA Journal*, 724, 1–114.

EFSA (European Food Safety Authority), 2011a. Scientific Opinion on Flavouring Group Evaluation 218, Revision 1 (FGE.218Rev1): alpha, beta-Unsaturated aldehydes and precursors from subgroup 4.2 of FGE.19: Furfural derivatives. *EFSA Journal* 9(3), 1840.

EFSA (European Food Safety Authority), 2011b. EFSA Panel on Dietetic Products, Nutrition and Allergies (NDA), Scientific Opinion on the substantiation of health claims related to caffeine and increase in physical performance during short-term high-intensity exercise (ID 737, 1486, 1489), increase in endurance performance (ID 737, 1486), increase in endurance capacity (ID 1488) and reduction in the rated perceived exertion/effort during exercise (ID 1488, 1490) pursuant to Article 13(1) of Regulation (EC) No 1924/20061. *EFSA Journal* 9(4), 2053.

EFSA (European Food Safety Authority), 2011c. EFSA Panel on Dietetic Products, Nutrition and Allergies (NDA), Scientific Opinion on the substantiation of health claims related to caffeine and increased fat oxidation leading to a reduction in body fat mass (ID 735, 1484), increased energy expenditure leading to a reduction in body weight (ID 1487), increased alertness (ID 736, 1101, 1187, 1485, 1491, 2063, 2103) and increased attention (ID 736, 1485, 1491, 2375) pursuant to Article 13 (1) of Regulation (EC) No 1924/20061. *EFSA Journal* 9(4), 2054.

EFSA (European Food Safety Authority), 2015a. EFSA panel on dietetic products, Nutrition and Allergies (NDA), scientific opinion on the safety of caffeine. *EFSA Journal* 13(5), 4102.

EFSA (European Food Safety Authority), 2015b. EFSA panel on contaminants in the food chain (CONTAM) scientific opinion on acrylamides in food. *EFSA Journal* 13(6), 4104.

FDA (Food and Drug Administration), 2016. *FDA Guidance for Industry Acrylamide in Foods*. FDA Office of Food Safety, USA.

Farah, A. and Donangelo, C., 2006. Phenolic compounds in coffee. *Brazilian Journal of Plant Physiology* 18(1), 23–36.

Farah, A., 2012. Coffee constituents. In: Chu, Y.-F. (Ed.), *Coffee: Emerging Health Effects and Disease Prevention*. IFT Press/Willey-Blackwell, USA, pp. 21–58.

Fehlau, R. and Netter, K. J.1990. Effect of untreated and non-irritating purified coffee and carbonic acid hydrytryptamides on the gastric mucosa in the rat Z. *Gastroenterol*, 28, 234–8.

Feldman, M. and Barnett, C., 1995. Relationships between the acidity and osmolarity of popular beverages and reported postprandial heartburn. *Gastroenterology*, 108, 125–31.

Feng, R., Lu, Y., Bowman, L. L., Qian, Y., Castranova, V. and Ding, M., 2005. Inhibition of activator Protein-1, NF-κB, and MAPKs and induction of phase 2 detoxifying enzyme activity by chlorogenic acid. *Journal of Biological Chemistry* 280, 27888–95.

Ferreira, I. M. P. L. V. O., Pinho, O. and Petisca, C., 2017. Potential effects of furan and related compounds on health. In: *Coffee: Chemistry, Quality and Health Implications* (Ed.: Farah, A.). Royal Society of Chemistry, London, UK. In press.

Floegel, A., Pischon, T., Bergmann, M. M., Teucher, B., Kaaks, R, Boeing, H., 2012. Coffee consumption and risk of chronic disease in the European Prospective Investigation into Cancer and Nutrition (EPIC)–Germany study. *American Journal Clinical Nutrition* 95, 901–8.

Fogliano, V. and Morales, F. J., 2011. Estimation of dietary intake of melanoidins from coffee and bread. *Food and Function* 2, 117–23.

Folmer, B., Farah, A., Fogliano, V. and Jones, L. 2017. Human wellbeing – Sociability, performance and health. In: *The Craft and Science of Coffee*, 1st Ed. (Ed.: Folmer, B.). Elsevier, London, UK, pp. 493–510.

Freedman, N., Park, Y., Abnet, C. C., Hollenbeck, A. R. and Sinha, R., 2012. Association of coffee drinking with total and cause-specific mortality. *New England Journal of Medicine* 366(20), 1891–904.

Garfinkel, B. D., Webster, C. D. and Sloman, L., 1981. Responses to methylphenidate and various does of caffeine in children with attention deficit disorder. *The Canadian Journal of Psychiatry/ La Revue canadienne de psychiatrie* 26(6), 395–401.

Garsetti, M., Pellegrini, N., Baggio, C. and Brighenti, F., 2000. Antioxidant activity in human faeces. *British Journal of Nutrition* 84(5), 705–10.

Gebicki, J., Marcinek, A., Chlopicki S. and Adamus, J. 2008. The use of quaternary pyridinium compounds for vasoprotection and/or hepatoprotection. Patent WO2008104920A1.

Gerson, M., 1978. The cure of advanced cancer by diet therapy: a summary for 30 years of clinical experimentation. *Physiological Chemistry and Physics* 10, 449–64.

Glória and Engeseth , 2017a. Potential beneficial effects of bioactive amines on health. In:*Coffee: Chemistry, Quality, and Health Implications* (Ed.: Farah, A.). Royal Society of Chemistry, London, UK. In press.

Glória and Engeseth , 2017b. Potential adverse effects of coffee bioactive amines to human health. In *Coffee: Chemistry, Quality, and Health Implications* (Ed.: Farah, A.). Royal Society of Chemistry, London, UK. In press.

Gniechwitz, D., Brueckel, B., Reichardt, N., Blaut, M., Steinhart, H. and Bunzel, M. 2007. Coffee dietary fiber contents and structural characteristics as influenced by coffee type and technological and brewing procedures. *Journal of Agricultural and Food Chemistry* 55, 11027–34.

Gniechwitz, D., Reichardt, N., Ralph, J., Blaut, M., Steinhart, H.and Bunzel, M., 2008. Isolation and characterisation of a coffee melanoidin fraction. *Journal of the Science of Food Agriculture* 88, 2153–60.

Goldstein, E. R., Ziegenfuss, T., Kalman, D., Kreider, R., Campbell, B., Wilborn, C., Taylor, L., Willoughby, D., Stout, J., Graves, B. S., Wildman, R., Ivy, J. L., Spano, M., Smith, S. E. and Antonio, J., 2010. International society of sports nutrition position stand: Caffeine and performance. *Journal of the International Society of Sports Nutrition* 7, 5.

Grembecka, M., Malinowska, E. and Szefer, P., 2007. Differentiation of market coffee and its infusions in view of their mineral composition. *Science of the Total Environment* 383, 59–69.

Gross, G., Jaccaud, E. and Hugget, A.C., 1997. Analysis of the content of the diterpenes cafestol and kahweol in coffee brews. *Food and Chemical Toxicology* 35, 547–54.

Health Canada, 2011. Information for Parents on Caffeine in Energy Drinks.www.hc-sc.gc.ca/fn-an/securit/addit/caf/faq-eng.php, Accessed January 2017.

Hedden, T. and Gabrieli, J. D. E., 2004. Insights into the ageing mind: A view from cognitive neuroscience. *Nature Reviews Neuroscience* 5, 87–97.

Higdon, J. V. and Frei, B., 2006. Coffee and health: A review of recent human research. *Critical Reviews in Food Science and Nutrition* 46, 101–23.

Hirakawa, N., Okauchi, R., Miura, Y. and Yagasaki, K., 2005. Anti-invasive activity of niacin and trigonelline against cancer cells. *Bioscience, Biotechnology and Biochemistry* 69(3), 653–8.

Hogervost, E., Bandelow, S., Schmitt, J., Jentjens, R., Oliveira, M., Allgrove, J., Carter, T. and Gleeson, M., 2008. Caffeine improves physical and cognitive performance during exhaustive exercise. *Medicine and Science in Sports and Exercise* 40, 1841–51.

Hong, B. N., Yi, T. H., Park, R., Kim, S. Y. and Kang, T. H., 2008. Coffee improves auditory neuropathy in diabetic mice. *Neuroscience Letters* 441(3), 302–6.

Huang, Z., Urade, Y. and Hayaishi, O., 2011. The role of adenosine in the regulation of sleep. *Current Topics in Medicinal Chemistry* 11, 1047–57.

International Agency for Research on Cancer (IARC), 2010. Some non-heterocyclic polycyclic aromatic hydrocarbons and some related exposures. IARC Monogr Eval Carcinog Ris 1997; USDA National Nutrient Database for Standard Reference, release 28, 2015 any Maryland, USA

International Coffee Organization, 2014. World coffee trade (1923–2013): A review of the markets, challenges and opportunities facing the sector. 112thSession.http://www.ico.org/news/icc-111-5-r1e-world-coffee-outlook.pdf

International Diabetes Federation (IDF), 2012. *Clinical Guidelines Task Force Global Guideline for Type 2 Diabetes.*

International Diabetes Federation (IDF), 2015. *Diabetes Atlas*, 5th Ed. Brussels, Belgium.

Islami, F., Boffetta, P., Ren, J. S., Pedoeim, L., Khatib, D. and Kamangar, F., 2009. High-temperature beverages and foods and esophageal cancer risk – A systematic review. *International Journal of Cancer* 125(3), 491–524.

Jee, S. H., He, J., Appel, L. J., Whelton, P. K., Suh, I. and Klag, M. J., 2001. Coffee consumption and serum lipids: A meta-analysis of randomized controlled clinical trials. *American Journal of Epidemiology* 153, 353–62.

Jenkins, N. T., Trilk, J. L., Singhal, A., O'Connor, P. J. and Cureton, K. J., 2008. Ergogenic effects of low doses of caffeine on cycling performance. *International Journal of Sport Nutrition and Exercise Metabolism* 18(3), 328–42.

Ji L., Jiang, P., Lu, B., Sheng, Y., Wang, X. and Wang, Z., 2013. Chlorogenic acid, a dietary polyphenol, protects acetaminophen-induced liver injury and its mechanism. *Journal of Nutrition and Biochemistry* 24(11), 1911–19.

Jurkowska, R. Z., Jurkowski, T. P. and Jeltsch, A., 2011. Structure and function of mammalian DNA methyltransferases. *European Journal of Chemical Biology* 12, 206–22.

Kalaska, B., Piotrowski, L., Leszczynska, A., Michalowsk, B., Kramkowski, K., Kaminski, T., Adamus, J., Marcinek A., Gebicki, J., Mogielnicki A. and Buczko, W., 2014. Antithrombotic effects of pyridinium compounds formed from trigonelline upon coffee roasting. *Journal of Agricultural and Food Chemistry*, 62(13), 2853–60.

Kasai H., Fukada, S., Yamaizumi, Z., Sugie, S. and Mori, H. 2000. Action of chlorogenic acid in vegetables and fruits as an inhibitor of 8-hydroxydeoxyguanosine formation in vitro and in a rat carcinogenesis model. *Food and Chemical Toxicology* 38, 467–71.

Kempf, K., Herder, C., Erlund, I., Kolb, H., Martin, S., Carstensen, M., Koenig, W., Sandwall, J., Bidel, S., Kuha, S. and Tuomilehto, J., 2010. Effects of coffee consumption on subclinical inflammation and other risk factors for type 2 diabetes: A clinical trial. *American Journal of Clinical Nutrition* 91, 950–7.

Khaw, K. T., Wareham, N., Bingham, S., Welch, A., Luben, R. and Day, N., 2008. Combined impact of health behaviours and mortality in men and women: the EPIC-Norfolk prospective population study. *PLoS Medicine* 5(1), e12.

Kuczkowski, K. M., 2009. Caffeine in pregnancy.*Archives of Gynecology and Obstetrics* 280, 695–8.

Kuwana, Y., Shiozaki, S., Kanda, T., Kurokawa, M., Koga, K., Ochi, M., Ikeda, K., Kase, H., Jackson, M. J., Smith, L. A., Pearce, R. K. and Jenner, P. G.1999. Antiparkinsonian activity of adenosine A2A antagonists in experimental models. *Advances in Neurology*, 80, 121–3.

Lachenmeier, D.W., 2015. Furan in coffee products: A probabilistic exposure estimation. In:*Coffee in Health andDiseasePrevention* (Ed.:Preedy, V.). Elsevier, 1st edition. New York and London, pp. 887–93.

Lakey-Beitia, J., Berrocal, R., Rao, K. S.and Durant, A. A., 2015. Polyphenols as therapeutic molecules in Alzheimer's disease through modulating amyloid pathways. *Molecular Neurobiology* 51(2), 466–79.

Lang, R., Wahl, A., Stark, T. and Hofmann, T.2011. Urinary N-methylpyridinium and trigonelline as candidate dietary biomarkers of coffee consumption. *Molecular Nutrition & Food Research* 5(11), 1613–23.

Lang, R., Bardelmeier, I., Weiss, C., Rubach, M., Somoza, V. and Hofmann, T., 2010. Quantitation of $^{\beta}$N-Alkanoyl-5-hydroxytryptamides in coffee by means of LC-MS/MS-SIDA and assessment of their gastric acid secretion potential using the HGT-1 cell assay. *Journal of Agricultural and Food Chemistry* 58(3), 1593–602.

Larsson, S. C. and Wolk, A., 2007. Coffee consumption and risk of liver cancer: A meta-analysis. *Gastroenterology* 132, 1740–5.

Larsson S. C. and Orsini, N., 2011. Coffee consumption and risk of stroke: a dose-response meta-analysis of prospective studies. *American Journal of Epidemiology* 174(9), 993–1001

Le Bloch J. L. V., Chetiveaux M., Freuchet B., Magot T., Krempf M., Nguyen P. and Ouguerram K. 2010. Nicotinic acid decreases apolipoprotein B100-containing lipoprotein levels by reducing hepatic very low density lipoprotein secretion through a possible diacylglycerol acyltransferase 2 inhibition in obese dogs. *The journal of Pharmacology and Experimental Therapeutics*, 334, 583–89.

Lecoultre, V., Carrel, G., Egli, L., Binnert, C., Boss, A., MacMillan, E. L., Kreis, R., Boesch, C., Darimont, C. and Tappy, L., 2014. Coffee consumption attenuates short-term fructose-induced liver insulin resistance in healthy men. *American Journal of Clinical Nutrition* 99(2), 268–75.

Lipworth, L., Sonderman, J. S., Tarone, R. E. and McLaughlin, J., 2012. Review of epidemiologic studies of dietary acrylamide intake and the risk of cancer. *European Journal of Cancer Prevention* 21, 375–86.

Loomis, D., Kathryn, Z., Grosse, G. Y., Lauby-Secretan, B., El Ghissassi, F., Bouvard, V.,et al. 2016. Carcinogenicity of drinking coffee, mate, and very hot beverages.*The Lancet Oncology* 17(7),877–8.

Lopez-Garcia, E., Rodriguez-Artalejo, F., Rexrode, K. M., Logroscino, G., Hu, F. B. and van Dam, R. M., 2009. Coffee consumption and risk of stroke in women. *Circulation* 119(8),1116–23.

Ludwig, I. A., Clifford, M. N., Lean, M. E. J., Ashihara, H. and Crozier, A., 2014. Coffee: Biochemistry and potential impact on health.*Food and Function* 5, 1695–717.

Machado, S. R., Parise, E. R. and Carvalho, L., 2014. Coffee has hepatoprotective benefits in Brazilian patients with chronic hepatitis C even in lower daily consumption than in American and European populations. *Brazilian Journal of Infectious Disease* 18(2), 170–6.

Macrae, R., 1985. Nitrogenous compounds. In: *Coffee. Volume 1 Chemistry*, 1st Ed. (Ed.:Clarke, R. J.and Macrae, R.). Elsevier, London and New York, p.115.

Malerba, S., Turati, F., Galeone, C., Pelucchi, C., Verga, F., La Vecchia, C. and Tavani, A., 2013. A meta-analysis of prospective studies of coffee consumption and mortality for all causes, cancers and cardiovascular diseases.*European Journal of Epidemiology* 28(7), 527–39.

Moreira, A., Coimbra, M., Nunes, F. M., Passos, C. P., Santos, S. A., Silvestre, A. J., Silva, A., Rangel, M.and Domingues, M. R. M., 2015. Chlorogenic acid-arabinose hybrid domains in coffee melanoidins: Evidences from a model system. *Food Chemistry* 185, 135–44.

Moura-Nunes, N., Perrone, D., Farah, A. and Donangelo, C., 2009. The increase in human plasma antioxidant capacity after acute coffee intake is not associated with endogenous non-enzymatic antioxidant components. *International Journal of Food Sciences and Nutrition* 60(supp 6), 173–81.

Mucci, L. A. and Adami, H. O., 2009. The Plight of the Potato: Is dietary acrylamide a risk factor for human cancer?*Journal of National Cancer Institute* 101(9), 618–21.

Navarini, L., Colomban, S., Caprioli, G.and Sagratini, G., 2017. Isoflavones, Lignans and other minor polyphenols. In:*Coffee: Chemistry, Quality and Health Implications* (Ed. Farah, A.). Royal Society of Chemistry, London, UK. In press.

Nathan, J., Panjwani, S., Mohan, V., Joshi, V. and Thakurdesai, P. A., 2014. Efficacy and safety of standardized extract of Trigonella foenum-graecum L seeds in an adjuvant to L-Dopa in the management of patients with Parkinson's disease. *Phytotherapy Research* 28(2), 172–8.

Nawrot, P., Jordan, S., Eastwood, J., Rotstein, J, Hugenholtz, A. and Feeley, M., 2003. Effects of caffeine on human health. *Food Additives and Contaminants* 20, 1–30.

Nehlig A. and Debry, G., 1994. Consequences on the newborn of chronic maternal consumption of coffee during gestation and lactation: A review. *Journal of the American College of Nutrition* 13, 6–21.

Nehlig, A., 2010. Is caffeine a cognitive enhancer? *Journal of Alzheimer's Disease* 20(S1), 85–94.

Nkondjock, A., 2012. Coffee and cancers. In:*Coffee: Emerging Health Effects and Disease Prevention* (Ed.: Chu, Y.-F.). 1st edition, Wiley-Blackwell, Oxford, UK, pp. 293–306.

Nunes F. M., Coimbra M. A., Duarte A. C., and Delgadillo I., 1997. Foamability, foam stability, and chemical composition of espresso coffee as affected by the degree of roast, *J. Agric. Food Chem.*, 45 (8), 3238–43.

Oboh, G., Agunloye, O. M., Akinyemi, A. J., Ademiluyi, A. O. and Adefegha, S. A., 2013. Comparative study on the inhibitory effect of caffeic and chlorogenic acids on key enzymes linked to Alzheimer's disease and some pro-oxidant induced oxidative stress in rats' brain-in vitro. *Neurochemistry Research* 38(2), 413–19.

Obuleso, M., Dowlathabad, M. R. and Bramhachari, P. V., 2011. Carotenoids and Alzheimer's Disease: An insight into therapeutic role of retinoids in animal models. *Neurochemistry International* 59(5), 535–41.

O'Connor, P. J., Motl, R. W., Broglio, S. P. and Ely, M. R., 2004. Dose-dependent effect of caffeine on reducing leg muscle pain during cycling exercise is unrelated to systolic blood pressure. *Pain* 109, 291–8.

Oliveira, M., Casal, S., Morais, S., Alves, C., Dias, F., Ramos, S., Mendes, E., Delerue-Matos, C. and Oliveira, B. P. P., 2012. Intra- and interspecific mineral composition variability of commercial coffees and coffee substitutes. Contribution to mineral intake. *Food Chemistry* 130, 702–9.

Oliveira, M., Ramos, S., Delerue-Matos, C. and Morais S, 2015. Espresso beverages of pure origin coffee: Mineral characterization, contribution for mineral intake and geographical discrimination. *Food Chemistry* 177, 330–8.

Olson, C. A., Thornton, J. A., Adam, G. E. and Lieberman, H. R., 2010. Effects of 2 adenosine antagonists, quercetin and cafeïne, on vigilance and mood. *Journal of Clinical Psychopharmacology* 30(5), 573–8.

Passos, C. P., Cepeda, M. R., Ferreira, S. S., Nunes, F. M., Evtuguin, D. V., Madureira, P., Vilanova, M.and Coimbra, M. A., 2014. Influence of molecular weight on in vitro immunostimulatory properties of instant coffee. *Food Chemistry* 161, 60–6.

Petraco, M., 2005. The cup. In: *Espresso Coffee: The Science of Quality*, 2nd edition (Eds: Illy, A.and Viani, R.). Elsevier Academic Press, Italy, pp. 290–8.

Quan H.Y., Kim D.Y., and Chung S.H, 2013. Caffeine attenuates lipid accumulation via activation of AMP-activated protein kinase signaling pathway in HepG2 cells *BMB Reports* 46(4), 207–12.

Ramassamy, C., 2006. Emerging role of polyphenolic compounds in the treatment of neurodegenerative diseases: A review of their intracellular targets. *European Journal of Pharmacology* 545, 51–64.

Ramos, S., 2008. Cancer chemoprevention and chemotherapy: Dietary polyphenols and signaling pathways. *Molecular Nutrition and Food Research*, 52, 507–26.

Rebello, S.A. and Van Dam, R., 2013. Coffee consumption and cardiovascular health: Getting to the heart of the matter. *Current Cardiology Reports* 15(10), 403–5.

Riedel, A., Hochkogler C. M., Lang, R., Bytof, G., Lantz, I., Hofmann, T. and Somoza, V., 2014. N-methylpyridinium, a degradation product of trigonelline upon coffee roasting, stimulates respiratory activity and promotes glucose utilization in HepG2 cells. *Food &Function* 5(3):454–62.

Rios, J. L., Francini, F. and Schinella, G. R., 2015. Natural products for the treatment of Type 2 diabete mellitus. *Planta Medica* 81(12–13), 975–94.

Ritchie, K., Carriere, I., de Mendonca, A., Portet, F., Dartigues, J. F., Rouaud, O., Barberger-Gateau, P. and Ancelin, M. L., 2007. The neuroprotective effects of caffeine. A prospective population study. *Neurology* 69, 536–45.

Rodrigues, D.A.C.and Casal, S., 2017. β-Carbolines. In: *Coffee, Chemistry, Quality and Health Implications* (Ed.: Farah, A.). Royal Society of Chemistry, London, UK. In Press.

Rosso, A., Mossey, J. and Lippa, C. F., 2008. Review: Caffeine: Neuroprotective functions in cognition and alzheimer's disease. *American Journal of Alzheimer's Disease and Other Dementias* 23(5), 417–22.

Rubach M., Lang R., Bytof G., Stiebitz H., Lantz I., Hofmann T. and Somoza V. 2014. A dark brown roast coffee blend is less effective at stimulating gastric acid secretion in healthy volunteers compared to a medium roast market blend. *Mol Nutr Food Res.*, 58(6), 1370–3.

Rufián-Henares, J.A. and Morales, F., 2007. Functional Properties of melanoidins: In vitro antioxidant, antimicrobial and antihypertensive activities. *Food Research International*, 40(8), 995–1002.

Ryan, L., Hatfield, C. and Hofstetter, M., 2002. Caffeine reduces time-of-day effects on memory performance in older adults. *Psychological Science*, 13(1), 68–71.

Saab, S., Mallam, D., Cox, G.A. and Tong, M.J., 2014. Impact of coffee on liver diseases: A systematic review. *Liver International*, 34(4), 495–504.

Sales, A., Miguel, M. A. and Farah, A., 2017. Potential prebiotic effect of coffee. In *Coffee: Chemistry Quality and Health Implications* (Ed.: Farah, A.). Royal Society of Chemistry, London, UK. In press.

Santos, M. D., Almeida, M. C., Lopes, N. P. and Souza, G. E. P. 2006, Evaluation of the anti-inflammatory, analgesic and antipyretic activities of the natural polyphenol chlorogenic acid. *Biological and Pharmaceutical Bulletin* 29, 2236–40.

Santos, C., Costa, J., Santos, J., Vaz-Carneiro, A. and Lunet, N., 2010. Caffeine intake and dementia: Systematic review and meta-analysis. *Journal of Alzheimer's Disease* 20(S1), 187–204.

Santos, I. S., Matijasevich, A. and Domingues, M.R., 2012. Maternal caffeine consumption and infant nighttime waking: Prospective cohort study. *Pediatrics* 129, 860–8.

Satel, S., 2006. Is caffeine addictive? – a review of the literature. *American Journal of Drug and Alcohol Abuse* 32 (4), 493–502.

Sengpiel. V., Elind, E., Bacelis, J., Nilsson, S., Grove, J., Myhre, R., Haugen, M., Meltzer, H. M., Alexander, T., Jacobsson, B. and Brantsaeter, A. L., 2013. Maternal caffeine intake during pregnancy is associated with birth weight but not with gestational length: results from a large prospective observational cohort study. *BioMed Central Medicine* 11, 42–60.

Smith, A., Sutherland, D. and Christopher, G., 2005. Effects of repeated doses of caffeïne and mood and performance of alert and fatigued volunteers. *Journal of Psychopharmacology* 19(6), 620–6.

Smith, R. F., 1987. A History of coffee. In:*Coffee: Botany, Biochemistry and Production of Beans and Beverage* (Clifford, M. N.and Wilson, K. C. (Eds)). 1st edition, Croom Helm, Ney York, pp. 1–12.

Somoza V., Lindenmeier M., Wenzel E., Frank O., Erbersdobler H. F., and Hofmann T. (2003). Activity-guided identification of a chemopreventive compound in coffee beverage using in vitro and in vivo techniques. *J. Agric. Food Chem.*, 51 (23), 6861–9.

Speer, K. and Kölling-Speer, I.2017. Lipids. In: *Coffee: Chemistry, Quality, and Health Implications* (Ed.: Farah, A.).Royal Society of Chemistry, London, UK. In press.

Spriet, L. L. and Gibala, M. J., 2004. Nutritional strategies to influence adaptations to training.*Journal of Sports Sciences* 22, 127–41.

Stavchansky, S., Combs, A., Sagraves, R., Delgado, M. and Joshi, A., 1988. Pharmacokinetics of caffeine in breast milk and plasma after single oral administration of caffeine to lactating mothers. *Biopharmaceutics and Drug Disposition* 9, 285–99.

Terry, P., Lagergren, J., Wolk, A. and Nyren, O., 2000. Reflux-inducing dietary factors and risk of adenocarcinoma of the esophagus and gastric cardia. *Nutrition and Cancer* 38(2), 186–91.

Tohda, C., Kuboyama, T. and Komatsu, K., 2005. Search for natural products related to regeneration of the neuronal network. *Neurosignals* 14(1–2), 34–45.

Torres, T. and Farah, A., 2016. Coffee, maté, açaí and beans are the main contributors to the antioxidant capacity of Brazilian's diet. *European Journal of Nutrition* March 14. 56(4):1523–33. doi:10.1007/s00394-016-1198-9.

Trugo, L. C., 1985. Carbohydrates. In: *Coffee. Volume 1 Chemistry*, 1st Ed. (Ed.: Clarke, R. J.and Macare, R.).Elsevier, London and New York, p.83.

Ukers, W. H., 1935. *All About Coffee*, 2nd Ed. Tea and Coffee Trade Journal Co., New York.

United States Department of Agriculture (USDA), 2015. USDA Scientific Report of the Dietary Guidelines Advisory Committee.

Urgert, R., van der Weg, G., Kosmeijer-Schuil, T. G., van de Bovenkamp, P., Hovenier, R. and Katan, M. B., 1995. Levels of the cholesterol-elevating diterpenes cafestol and kahweol in various coffee brews.*Journal of Agricultural and Food Chemistry* 43(8), 2167–72.

Urgert, R. and Katan, M. B., 1997. The cholesterol-raising factor from coffee beans. *Annual Review of Nutrition* 17, 305–24.

USDA National Nutrient Database for Standard Reference (Release 28, released September 2015, slightly revised May 2016) United States Department of Agriculture, https://ndb.nal.usda.gov/ndb/.

USDA National Nutrient Database, 2017. United States Department of Agriculture, Agricultural Research Service, USDA Food Composition Databases, https://ndb.nal.usda.gov/ndb/, Accessed in February 13 2017.

van Boxtel, M. P. J. and Schmitt, J. A. J., 2004. Age related changes in the effects of caffeine on memory and cognitive performance. In: *Coffee, Tea, Chocolate and the Brain* (Ed.: Nehlig, A.).CRC Press, Boca Raton, Florida, pp. 85–97.

van Dijk A.E., Olthof, M. R., Meeuse, J. C., Seebus, E., Heine, R. J. and van Dam, R. M., 2009. Acute effects of decaffeinated coffee and the major coffee components chlorogenic acid and trigonelline on glucose tolerance.*Diabetes Care* 32, 1023–5.

Vitaglione, P., Fogliano, V. and Pellegrini, N., 2012. Coffee, colon function and colorectal cancer. *Food and Function* 3, 916–22.

Waite, M., 2015. Treatment for Alzheimer's disease: Has anything changed? *Australian Prescriber* 38, 60–3.

Wang W., Basinger A., Neese R. A., Shane B., Myong S. A., Christiansen M. and Hellerstein M. K. 2001. Am J Effect of nicotinic acid administration on hepatic very low density lipoprotein-triglyceride production. *Physiol Endocrinol Metab.* 280, 3, E540–7.

WHO/FAO (World Health Organization and Food and Agriculture Organization of the United Nations), 2002. Human vitamin and mineral requirements. Report of a Joint FAO/WHO Expert Consultation, Bangkok, Thailand.FAO, Rome. http://www.fao.org/docrep/004/y2809e/y2809e00.htm

WHO 2017, http://www.who.int/mediacentre/factsheets/fs310/en/accessed in January 2017.

Yamamoto, T., Yoshizawa, K., Kubo, S., Emoto, Y., Hara, K., Waters, B., Umehara, T., Murase, T. and Ikematsu, K., 2015. Autopsy report for a caffeine intoxication case and review of the current literature. *Journal of Toxicologic Pathology* 1, 33–6.

Yoshinari, O. and Igarashi, K., 2010. Anti-diabetic effect of trigonelline and nicotinic acid, on KK-Ay mice. *Current Medicinal Chemistry* 17(20), 2196–202.

Youngberg, M. R., Karpov, I. O., Begley, A., Pollock, B. G. and Buysse, D. J., 2011. Clinical and physiological correlates of caffeine and caffeine metabolites in primary insomnia. *Journal of Clinical Sleep Medicine* 7, 196–203.

Chapter 3

Instrumentation and methodology for the quantification of phytochemicals in tea

Ting Zhang, China University of Geosciences and Huanggang Normal University, China; Xiaojian Lv, Yin Xu, Lanying Xu and Tao Long, Huanggang Normal University, China; Chi-Tang Ho, Rutgers University, USA; and Shiming Li, Huanggang Normal University, China and Rutgers University, USA

1 Introduction

In addition to water, tea (*Camellia sinensis*) is the most popular beverage in the world. China is the world's largest producer of tea, contributing to 36% of the total global tea production, followed by India (21.2%), Kenya (7.8%), Sri Lanka (7.0%), Turkey (4.8%), Vietnam (4.6%) and Iran (3.3%). Tea contains compounds such as polyphenols, amino acids, vitamins, carbohydrates and purine alkaloids such as caffeine which possess important physiological properties and health-promoting benefits.[1] Tea is one of the world's most popular non-alcoholic flavoured beverages. Tea and its extracts have been widely used in beverages, functional foods, cosmetics and other fields. With the increasing research on the functional components of tea, the extraction and screening technology of antioxidant activity from various effective components of teas have become a hotspot for the investigation of the tea.

http://dx.doi.org/10.19103/AS.2017.0036.16

The variety of tea found in a region is closely associated with the growing conditions, geographical origin and the approach used for plucking and processing of tea leaves. It can be classified into three types: unfermented green and white teas, partially fermented oolong tea and completely fermented black tea (Fig. 1).[2]

White tea is prepared from young tea leaves or buds covered with tiny and silvery hairs, harvested before being fully opened. The tea materials are dried immediately (Fig. 1) after picking in order to inhibit oxidation process, providing a light and delicate taste.[3] The tea buds, harvested only once a year in the early spring, are shielded from sunlight during growth in order to reduce the formation of chlorophyll, giving the young leaves a white appearance.[4]

Despite its simple process of drying without fermentation, it has been claimed to pose some health issues, as (i) it contains greater level of antioxidants; (ii) it contains lower caffeine content than any other tea from the *C. sinensis* plant (green, black or oolong); (iii) inhibits pancreatic lipase activity and anti-mutagenicity to a higher level than green tea or black tea. However, the available composition data showed that the first two claims are completely inaccurate, for which the concentration of catechins and corresponding antioxidant activities are not necessarily higher in white tea than in green or black tea,[5–7] and the same situation was found with regard to the concentration of total caffeine.[8] An *in vitro* experiment showed that white tea was equally efficient as green tea and black tea in inhibiting pancreatic lipase activity.[9] In addition, Salmonella assay has shown that white tea exhibits potent anti-mutagenic activity against several heterocyclic amines, such as 2-amino-1-methyl-6-phenylimidazo (4,5-b)-pyridine (PhIP).[10] Subsequently, white tea has been proved to inhibit the formation of PhIP-induced colonic aberrant crypts in rats,[11] the formation of intestinal polyps in ApcMin/+ mouse[12] and β-catenin/TCF-4 reporter activity in HEK293 cells.[13]

Green tea, prepared from freshly harvested tea leaves which are steamed to deactivate the polyphenol oxidase enzyme, and then rolled and dried (Fig. 1),[14] is an evergreen plant that grows primarily in tropical and temperate regions of Asia, especially China, India, Sri Lanka and Japan. It is also cultivated in several African and South American countries. The characteristic taste of green tea is made up of a mixture of bitterness, astringency, meaty (umami) taste, sweetness and slight sourness (Table 1). Its chemical composition is very similar to that of the fresh tea leaf,[15] which has abundant chemicals with maximum positive effects, such as antioxidant effect, antihypertensive effect, body weight control,

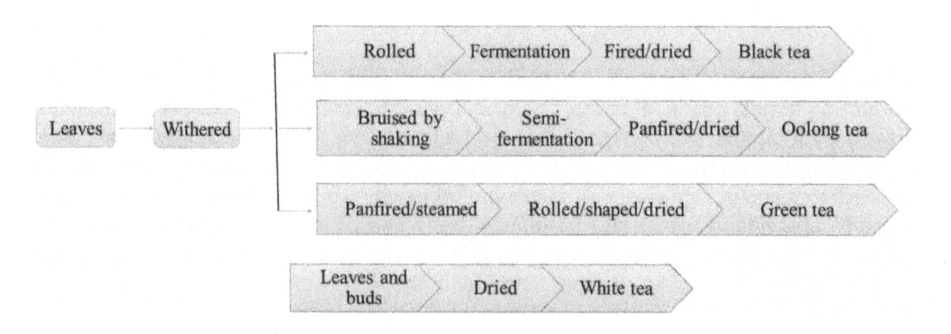

Figure 1 Manufacturing process of black, oolong, green and white tea.

Table 1 Main chemical compounds in green, black or oolong tea and their typical concentrations (mg/g, dry weight base)[16]

Principal components	Green tea	Black tea	Oolong tea
Catechins	60–150	10–20	50–70
Theaflavins (TFs)	ND	5–15	3–10
Thearubigins (TRs)	ND	60–150	40–110
Theabrownin (TB)	ND	40–90	~20
Theanine	15–20	8–14	10–20
Amino acid	30–40	25–35	30–35
Peptides/protein	130–270	130–160	130–160
Organic	10–15	10–20	18–25
Sugars	20–30	20–40	20–40
Other carbohydrates	180–270	150–260	170–270
Lipids	35–75	40–80	40–80
Caffeine	30–40	30–40	30–40
Other methylxanthines	2–2.5	2–2.5	2–2.5
Minerals/ash	~60	~60	~60
Chlorophyll	1–5	0.1–0.2	0.3–2
Volatile compounds	0.2–0.5	0.1–0.3	0.2–0.5
Colour	Greenish or yellowish-green without any trace of red or brown	Bright reddish and brown	Reddish-brown (moderate to heavy fermented); dark greenish (light fermented)
Taste	Strong astringency and bitterness, median umami and sweetness and slight sourness	Brisk and strong taste	Mellowness and sweetness, lower astringency and stronger sweetness than green tea
Aroma	Fresh flavour	Sweet and rich flavour	Fruity and floral

antibacterial and antiviral activity, bone mineral density increase, anti-fibrotic properties and neuroprotective power, on human beings.

Oolong tea is prepared from freshly harvested young green shoots which undergo a semi-fermentation process (Fig. 1). The final fermentation degree of oolong tea mainly depends on the taste of the consumers' preference and is usually in the range of 20–80%.[17] It has been reported that the mellowness and sweetness of oolong tea are attributed to the presence of non-oxidized catechins, thearubigins, some polyphenolic compounds, caffeine, free amino acids and related sugars.[18] The astringency of oolong tea is lower than that of green tea, while its sweetness is stronger than that of green tea. Oolong tea contains a high number of antioxidants,[19] which inhibits neurosphere adhesion, cell migration and neurite outgrowth in rat neurospheres.[20]

Black tea accounts for approximately 78% of the world's total tea production, and is mainly consumed in Western countries. Its production process includes withering, fermenting and finally drying (Fig. 1). During withering, the leaves take on a form facilitating the rolling process and when the cell structure of leaves is disrupted the fermentation process begins. In the fermentation process of black tea, about 75% of catechins contained in the tea leaves undergo enzymatic transformation (Fig. 2) consisting in oxidation and partial polymerization.[21] Although most of the EGCG antioxidants are oxidized during the fermenting process, black tea contains a high number of the antioxidant polyphenols such as flavonoids. The strong taste of black tea is attributed to its bitterness, astringency, sweetness, malty taste, green/grassy taste, 'caramel-like' and 'hay-like' characteristics.[22]

Many studies have demonstrated that black tea and its bioactive compounds have preventive effects on chronic inflammatory diseases (Table 2), carcinogenesis, cardiovascular diseases, obesity and metabolic syndrome, neurological disorders and longevity (Table 2).

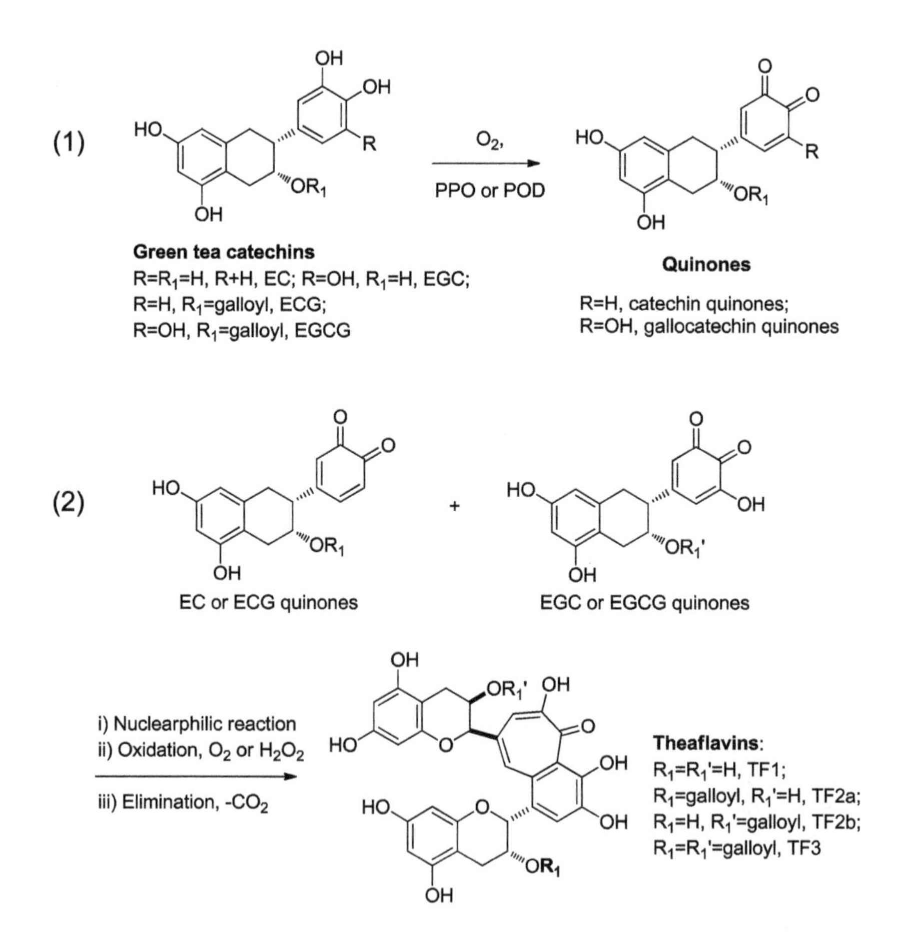

Figure 2 Enzymatic catalysed process of theaflavin formation.

Table 2 Effect of black tea compounds on chronic inflammatory diseases, anticancer property, prevention of cardiovascular diseases, obesity, metabolic syndrome and neurological disorders

Anti-inflammatory mechanisms		**References**
Black tea extract	Reduced lipopolysaccharide (LPS)-induced NO and O_2^- production as well as inducible nitric oxide synthase (iNOS) expression in murine macrophage	23
TF1	Suppressed tumour necrosis factor (TNF)-α-induced interleukin (IL)-8 gene expression through blockage of NF-κB activation and reduced DNA binding activity of activator protein (AP)-1	24
TF3	Inhibited inflammatory enzyme expression and cytokines production in murine macrophage	25
TF3	Suppressed TPA-induced ear oedema by inhibition of arachidonic acid metabolism	26
TF1	Inhibited LPS-induced adhesion molecule expression by suppression of NF-κB and c-Jun N-terminal kinases (JNK)	27
TF3	Provided protection against trinitrobenzene sulphonic acid (TNBS)–induced colitis and pro-inflammatory cytokines production in mice	28
TF2a	Decreased pro-inflammatory cytokines production in type IV allergy mice	29
TF3	Inhibited inflammatory cytokines production in human gingival fibroblasts	30, 31
Black tea extract	Helped maintain skeletal health through reduction of active osteoclasts, inflammatory cytokines production and oxidative stress	32
Black tea extract	Provided protection against high fat diet (HFD) fed non-alcoholic steatohepatitis (NASH)-induced bone skeletal dysfunctional changes in Wistar rats	33
TF3	Inhibited differentiation of osteoclasts through down-regulation of MMP-2 and MMP-9	34
Anticancer mechanisms		
Catechins	Reduced the concentration of 17beta-estradiol (E2) in women	35
Black tea	Inhibited 7,12-dimethylbenz(a)anthracene (DMBA)-induced skin tumorigenesis through activation of superoxide dismutase (SOD) and catalase (CAT) as well as induced apoptosis in mouse skin tumours	36
Black tea	Induction of apoptosis by tea polyphenols mediated through mitochondrial cell death pathway in mouse skin tumours	37
Black tea	Suppressed 1,2-dimethylhydrazine (DMH)-induced colonic tumorigenesis by inhibition of cyclin D1, c-myc and cyclooxygenase-2 (COX-2) gene expression through blockage of Wnt/β-catenin pathway	38

(Continued)

Table 2 (*Continued*)

Black tea extract	Provided protection against oxidative stress, interfered with the activity of carcinogen metabolizing enzymes and inhibited DMBA-induced tumour in hamster buccal pouch carcinogenesis model	39
TF1	Participated in the induction of apoptosis through increasing reactive oxygen species (ROS) production and activation of JNK-p38 mitogen-activated protein kinase (MAPK) signalling	40
TF1	Suppressed metalloproteinases (MMP)-2 gene expression and activity by down-regulation of epidermal growth factor receptor (EGFR) and NF-κB signalling	41
TF1	Induced apoptosis through Fas death receptor/caspase-8 cascade and inhibited pAkt/pBad survival pathway in p53-mutated human breast cancer cells	42
TF1	Inhibited gene and protein expression of fatty acid synthase (FAS) in MCF-7 breast cancer cells	43
TF1	Against lung carcinogenesis through induction of apoptosis and inhibition of COX-2 expression in mice	44, 45
TF1	Inhibited bronchiolar cell proliferation and tumour formation in 4-(methyl nitrosamino)-1-(3-pyridyl)-1-butanone (NNK)-induced lung carcinogenesis in A/J mice	46
TF2	Induced apoptosis in human colon cancer cell line and repressed 12-O-tetradecanoylphorbol-13-acetate (TPA)–induced ear oedema in mice	47
TF2	Inhibited cell proliferation, induced apoptosis and down-regulated COX-2 gene expression in cancer cells	48
TF2	Induced apoptosis through cytochrome c-medicated cascade in U937 cells	49
TF3	Inhibited cell proliferation through suppression of phosphorylated EGFR and platelet-derived growth factor (PDGF)	50
TF3	Repressed EGFR expression through protein endocytosis and degradation	51
TF3	Induced apoptosis by increasing oxidative stress in human oral squamous cells	52
TF3	Induced cell cycle arrest combined with ascorbic acid in human lung adenocarcinoma cells	53
Protective effects on cardiovascular diseases		
Catechins	Associated with reduced risk of cardiovascular disease	54, 55
TF1, TF2a, TF2b, TF3	Inhibited Cu^{2+}-mediated low-density-lipoprotein (LDL) oxidation *in vitro*	56
TF1	Attenuated atherosclerosis in apolipoprotein E (Apo/E)-knockout mice through reduction of ROS and inflammatory cytokines production	57

Black tea extract	Potent inhibitors of the acetyltransferase and 1-alkyl-2-acetyl-sn-glycero-3-phosphocholine (PAF) biosynthesis that inhibited platelet aggregation	58
Black tea extract	Protected tert-butyl hydroperoxide (t-BHP)-induced malondialdehyde (MDA) and dityrosine levels in human vein endothelial cells (HUVEC)	59
TF3	Increased nitric oxide (NO) production and endothelial NO synthase (eNOS) activity in bovine aortic endothelial cells	60
TF3	Inhibited plasminogen activator inhibitor type 1 (PAI-1) activity	61
Black tea extract	Reduced risk of coronary heart disease and myocardial infarction	55, 62
	Reduced risk of coronary heart disease and myocardial infarction	63
	Reversed endothelial dysfunction in patients with coronary artery disease	64
Prevention of obesity and metabolic syndrome		
Black tea	Reduced HFD-induced body weight gain, improved hyperglycemia and glucose intolerance via increasing the level of glucose transporter type 4 (GLUT4) in male C57BL/6J mice	65
Black tea extract	Suppressed HFD-induced body weight increase, adipose tissue mass and liver lipid content through inhibition of intestinal lipid absorption in C57BL/6N mice	66
	Decreased sucrose-rich diet-induced body weight gain and hypercholesterolemia in male SD rats	67
	Reduced HFD-induced NASH through regulation of pro-oxidant and antioxidant status, protected hepatocellular damage and against apoptosis in Wistar rats	68
TF1	Suppressed steatosis and steatotic liver ischaemia-reperfusion (I/R) injury through inhibiting ROS production, hepatocyte apoptosis and inflammatory cytokine expression in male C57BL/6 mice	69
TF2a	Reduced lipid and triglyceride levels in fatty acid-overload HepG2 cells through induction of AMP-activated protein kinase (AMPK) signalling	70
TF1	Inhibited gene and protein expression of FAS in MCF-7 breast cancer cells	43
TF2a, TF2b	Reduced the solubility of cholesterol into micelles	71, 72
TF3	Suppressed hypertriacylglycerolemia through inhibition of pancreatic lipase activity in rats	73

(*Continued*)

Table 2 *(Continued)*

Effects on neurological disorders and longevity		
Black tea extract	Inhibited lipid membrane destabilization induced by amyloid β peptide (Aβ) 42 peptides *in vitro*	74
TF2b, TF3	Potent inhibitor of PAI-1 in human plasma that may slow down progression of Alzheimer's disease	61, 75
TF3	Inhibited Aβ fibril formation via reduction of ThT signals and assembly of Aβ fibrils into nontoxic, spherical aggregates *in vitro*	76
Black tea	Reduced risk of Parkinson's disease in the Singapore Chinese Health Study	77
Black tea extract	Protected 6-hydroxydopamine (6-OHDA)-induced dopaminergic neuron damage, improved motor and neurochemical deficits in Wistar rats	78
TF1	Attenuated 1-methyl-4-phenyl-1, 2, 3, 6-tetrahydropyridine (MPTP/p)-induced apoptosis and neurodegeneration in C57BL/6 mice through increase of dopamine transporter (DAT) and reduction of apoptosis in the substantia nigra	79
Black tea extract	Increased life span of Drosophila in fat-diet system	80
TF2a, TF2b,TF3	Induced phosphorylation of longevity factor forkhead box protein O 1a (FOXO1a) *in vitro*	81

2 Phytochemicals in tea: bioactive compounds

The chemical composition of tea depends on the following factors: genetic strain, climatic conditions, soil growth altitude, horticultural practices, the plucking season and sorting (grading) of tea leaves. There are many chemical compounds in tea. The main chemical compounds present in dry tea leaves are listed in Fig. 3. This section will discuss bioactive compounds, and Section 3 will cover flavour and colour compounds.

2.1 Polyphenols

Tea polyphenols, also called tea tannin, is the general name given to polyphenols. It contains more than 30 types of phenolic compounds, mainly composed of catechins, flavonoids, anthocyanins and four types of phenolic acids. Studies have shown that tea polyphenols have antioxidant, free radical scavenging, hypotensive, hypolipidemic, anti-radiation and anticancer properties, as well as other biological activities and pharmacological effects.

The predominant flavonoids in green tea are catechins and their gallates, which include about 70% of the total tea polyphenols such as: (–)-epicatechin (EC), (–)-epigallocatechin (EGC), (–)-epicatechin gallate (ECG approximately) and (–)-epigallocatechin gallate (EGCG).

During the process of fermentation, polyphenols in green tea are oxidized and subsequently polymerized. For example, catechin in black tea is transformed into teaflavin-3-3′-digallate and thearubigin, with its content reduced by approximately

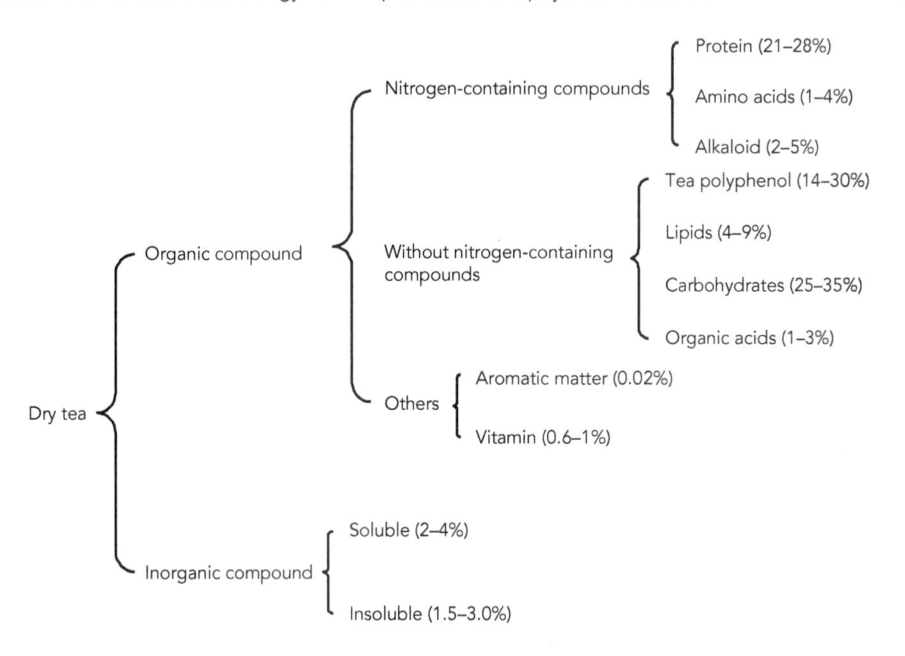

Figure 3 Main chemical compounds in dry tea leaves and their typical concentrations (dry weight base).

85%,[82] as shown in Table 1. The catechin content is higher in green tea than in black tea and oolong tea. Polyphenols in black tea include theaflavins (TFs in Table 3) and thearubigins (TRs). These are formed when catechin oxidizes during the catalysis of polyphenol oxidase and peroxidase in the fermentation process to manufacture black tea. Their particular structure produces yellowish and bright reddish colours and imparts a bright and yellowish appearance to the black tea brew. There are four major TFs (Fig. 4), namely, theaflavin (TF1; parent flavanols from EC and EGC); theaflavin-3-gallate (TF2a; from EC and EGCG); theaflavin-3′-gallate (TF2b; from ECG and EGC) and theaflavin-3, 3′-digallate (TF3; from ECG and EGCG). Due to their similar heterogeneity, solubility and chromatographic behaviours, the structures of TRs are still ambiguous. The TF content in black tea is usually less than 2.5%. It has been proven that polyphenols of black tea inhibit lipase activity and hence intestinal lipid absorption and suppress the increase of triglyceride levels in rat plasma.[83]

Oolong tea–polymerized polyphenols are produced by semi-fermentation and heating process which means that the oolong tea is partially oxidized with a lower concentration of polymeric polyphenols and higher concentrations of green tea catechins than black tea. (–)-Epigallocatechin 3-*O*-(3-*O*-methyl) gallate (EGCG3″Me, Fig. 5), the unique *O*-methylated form of EGCG that is present only in limited oolong and green teas, has been reported to possess potential prebiotic-like activity by modulating intestinal microbiota, contributing to the improvements of host health and oolong tea polyphenols, while EGCG and EGCG3″Me have been reported to possess impactful inhibitory activity *in vitro* and *in vivo*.[85]

Table 3 Nomenclatures of major tea catechins and theaflavins[84]

Polyphenol	Structure	No.	Name	R	R′	Acronym
Catechins		I	Epicatechin	H	H	EC
		II	Epigallocatechin	H	OH	EGC
		III	Epicatechin gallate	galloyl	H	ECG
		IV	Epigallocatechin gallate	galloyl	OH	EGCG
Theaflavins		V	Theaflavin	H	H	TF1
		VI	Theaflavin-3-monogallate	galloyl	H	TF2a
		VII	Theaflavin-3′-monogallate	H	galloyl	TF2b
		VIII	Theaflavin-3,3′-digallate	galloyl	galloyl	TF3

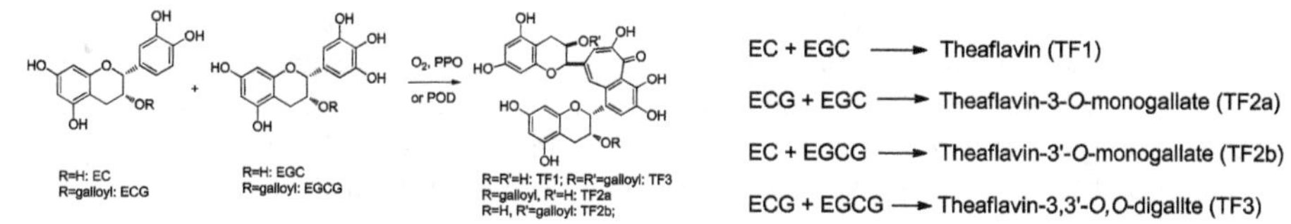

EC + EGC → Theaflavin (TF1)
ECG + EGC → Theaflavin-3-O-monogallate (TF2a)
EC + EGCG → Theaflavin-3'-O-monogallate (TF2b)
ECG + EGCG → Theaflavin-3,3'-O,O-digallite (TF3)

Figure 4 Major specific theaflavin formation from catechins.

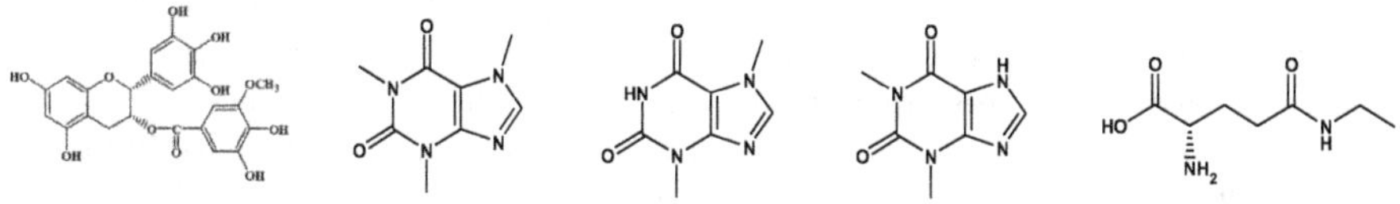

EGCG3''Me Caffeine Theobromine Theophylline L-Theanine

Figure 5 Structures of EGCG3"Me, alkaloids and L-theanine.

2.2 Alkaloids

The most abundant alkaloid component present in tea is purine. Caffeine is one of the major alkaloids in tea, and its content in dry tea leaves ranges from 2% to 5% (Fig. 3). Alkaloids (caffeine, theobromine and theophylline in Fig. 5) have a conjugated structure, which can absorb strong UV. Caffeine is easy to sublimate at a temperature above 100°C, and it has strong effects on the quality of tea, playing an important role in tea flavour, especially green tea. The bitter taste of a tea brew is often associated with a high caffeine content.

Generally, alkaloid undergoes some changes during tea product processing. Pure caffeine tastes unpleasant and bitter (threshold is approximately 3 mg/L in water). However, a part of caffeine can form complexes with catechins, theaflavins and thearubigins, which reduce the degree of bitterness.

2.3 Carbohydrates

Concentrations of total soluble carbohydrates in tea are one of the important physical and chemical indicators of the quality of tea. Tea contains a number of carbohydrates, which were formed through photosynthesis and metabolism. For example, free sugars in young tea shoots consist of mainly glucose, fructose, sucrose, raffinose and stachyose. Amylum content in young tea shoots is low. Pectin substances consist of galactose, arabinose, galacturonic acid, rhamnose and ribose. The carbohydrates in tea are mainly divided into three categories: monosaccharides, disaccharides and polysaccharides. Of these, monosaccharides and disaccharides are easily soluble in water, which are collectively referred to as water-soluble carbohydrates; the concentrations of monosaccharides and disaccharides directly affect the taste and quality of tea. The slight sweetness of tea may be partly attributed to the presence of sugars. Heating during the green tea manufacturing process can cause sugars to react with free amino acids to produce compounds from the Maillard reaction, with the roast aroma released.

Usually, tea polysaccharides (TPS) were found to be glycoconjugates in which a protein contains some carbohydrate chains covalently attached to a polypeptide backbone, such as via *O*- or *N*-linkages. Monosaccharides in TPSs are mainly composed of arabinose, galactose and glucose with the approximate molar ratio of 1:1:0.5. The protein part of TPSs is composed of 16 normal amino acids of which Gly, Glu, Val and Ala are in major proportion.

2.4 Amino acids

The concentrations of total free amino acids in tea range from 1% to 4% (Fig. 3). Tea is rich in free amino acids and contains more than 20 types of them when picked as young tea shoots. Amino acids are easily soluble in water and have a positive impact on tea infusion and flavour. Each free amino acid contributes a unique taste to tea such as bitter, sour, sweet or salty flavour. For example, free amino acids are the reason for the taste of mellowness in green tea. Among the amino acids present in tea, the content of theanine, glutamic acid, glutamine and arginine are the highest. Particularly, the highest content of amino acid is L-theanine (5-*N*-ethyl-glutamine, Fig. 5), usually called theanine, which accounts for more than 50% of the total amino acid content in tea leaves.[18] It is the main component responsible for the umami taste of tea and possesses some pharmacological activities.[86] For these reasons, theanine is usually used as an index for determining the characteristics and quality of tea. Moreover, it has been demonstrated that theanine can decrease the level of norepinephrine and serotonin in the brain, and that intake of theanine by hypertensive rats results in decreased blood pressure.[87] Theanine also has been illustrated to increase serotonin, dopamine and γ-aminobutyric acid (GABA) levels in the brain, which impart neuroprotective effects.[88,89]

Generally, free amino acids can be transformed to aldehydes by decarboxylation and deamination and to some volatile compounds of made tea. During the manufacturing process, free amino acids, especially L-arginine and L-theanine, may react with sugars to form furan, pyrazine and pyrrole under heated conditions, which contribute to the roast aroma of green tea. In addition, amino acids can undergo Strecker degradation to convert into carboxylic acids, aldehydes and alcohols when flavonoids and carbohydrates are also present.

2.5 Tea saponins

Saponins, a second group of metabolites mainly derived from plant materials, have been used extensively in drug-related applications due to their pharmaceutical properties. Tea saponin (TS) is extracted from tea seed, usually considered as residue and disposed of after the tea seed is extracted for oil. Saponins of tea leaf are different from those of tea seed. Some of the major tea leaf saponins are: sapogenins, barringtogenol C, camelliagenin A, R_1-barrigenol and A_1-barrigenol components of sugars such as glucuronic acid, galactose, xylose and arabinose. Acidic components such as angelic acid and cinnamic acid are not included in tea seed saponins.[90] Several saponins such as theasaponin B_1; assamsaponin J; isotheasaponin B_1-B_3; foliatheasaponin I, II, III, IV, V and floratheasaponin A have already been found to be present in green tea (Table 4).[90–93] Saponins or saponin-like substances have been proved to suppress methane production, reduce rumen protozoa counts and modulate fermentation patterns. Saponins are complex molecules consisting of non-sugar aglycones coupled to sugar chain units.

Table 4 Molecular formula for tea leaf saponins[90–93]

	FW	R_1	R_2	R_3	R_4	R_5
Theasaponin B_1	1306	(E)-Cinnzmoyl	Acetyl	H	Acetyl	Xylose
Assamsaponin J	1320	Acetyl	(E)-Cinnzmoyl	H	Acetyl	Rhamnose
Isotheasaponin B_1	1264	(E)-Cinnzmoyl	H	Acetyl	H	Xylose
Isotheasaponin B_2	1264	Acetyl	(E)-Cinnzmoyl		H	Xylose
Isotheasaponin B_3	1304	Angeloyl	(E)-Cinnzmoyl	H	H	Xylose
Foliatheasaponin I	1258	Tigloyl	Acetyl	H	Acetyl	Xylose
Foliatheasaponin II	1264	(E)-Cinnzmoyl	Acetyl	H	H	Xylose
Foliatheasaponin III	1258	Angeloyl	Acetyl	H	Acetyl	Xylose
Foliatheasaponin IV	1306	(Z)-Cinnzmoyl	Acetyl	H	Acetyl	Xylose
Floratheasaponin A	1216	Angeloyl	Acetyl	H	H	Xylose

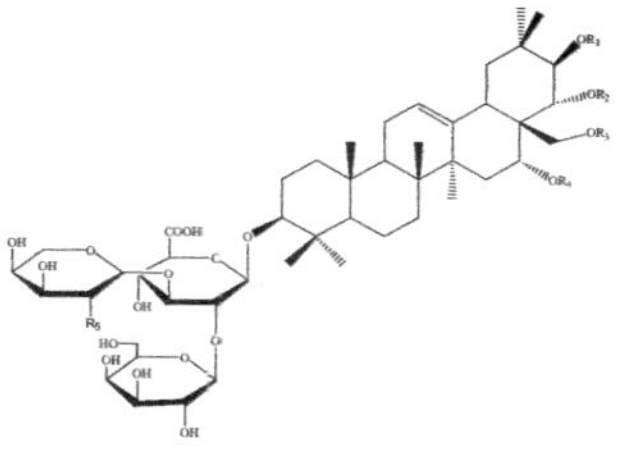

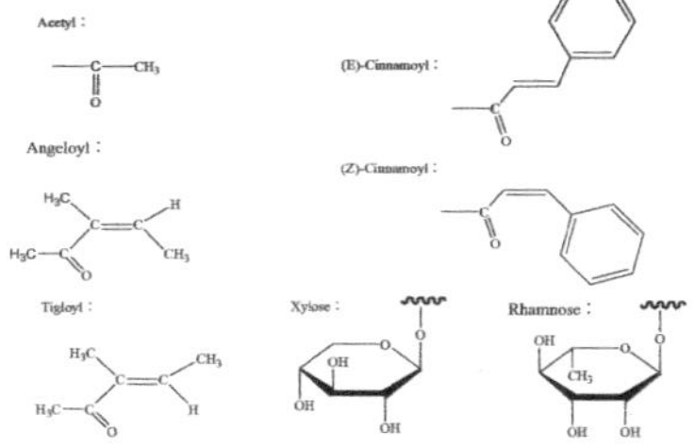

Acetyl :

Angeloyl :

Tigloyl :

(E)-Cinnamoyl :

(Z)-Cinnamoyl :

Xylose :

Rhamnose :

3 Phytochemicals in tea: flavour and colour compounds

3.1 The flavour compounds in tea

Volatile compounds

The market value of tea is highly dependent on its quality, which is determined by its colour, freshness, strength and aroma. More than 600 volatile compounds have been found to be present in tea leaves or tea beverages,[94] and the number of compounds is still found to increase as new compounds are being continuously discovered, with 41 of these compounds being identified as important contributors to the aroma of black tea.[93,94] These compounds are often classified into groups including hydrocarbons, alcohols, ketones, aldehydes, lactones, phenolic compounds, acids, esters, nitrogen and sulphur compounds and mixed oxygenated compounds. Some important tea volatile compounds, with their odour characteristics, are summarized in Table 5.

While phenolic compounds are responsible for the colour and the taste of tea, volatile compounds are fundamentals for tea odour and aroma which are the primary qualities found in fresh leaves. These volatile compounds are usually biosynthesized by plants. Secondary products are formed from substrates during production, for example, lipids, amino acids and carotenoids, formed through enzymatic and non-enzymatic reactions.

Table 5 Characteristics of flavour compounds in tea infusions[95,96]

	Volatile compounds	Odour threshold (μg/L)	Odour description
Fatty acid derived volatiles	Hexanol	92–97	Green, grassy
	Hexanal	10	Green, grassy, metallic
	Trans-2-Hexenal	190	Green, fruity
	Cis-3-Hexenol	13	Fresh, fruity, green
	Methyljasmonate	nd	Sweet, floral
Volatile terpenes	Linalool	0.6	Floral, citrus-like
	Linalooloxides	nd	Sweet, floral, citrus, fruity
	Geraniol	3.2	Floral, rose-like
Volatile phenylpropanoids/ benzenoids	2-phenylethanol	1000	Flowery, rose-like, honey-like
	Benzylalcohol	nd	Sweet, fruity
	Coumarin	2×10^{-5}	Sweet, camphoraceous
	Phenylacetaldehyde	nd	Honey-like
Carotenoid-derived volatiles	β-Ionone	0.2	Woody, violet-loke
	Damascenone	0.04	Honey-like, fruity

nd: not defined.

As a result, hundreds of volatile compounds have been found in black tea, with fewer numbers in oolong and green tea, due to the lesser degree of fermentation when manufacturing these varieties.

Different types of tea have different aromas and odours. The aroma of a particular tea is determined by the manufacturing process, variety of tea, developmental stages of fresh shoots, climatic conditions and agronomic management.[16] For instance, green tea contains fresh and greenish odour of (*Z*)-3-hexenol and its esters, and (*E*)-2-hexenal. Roasted teas (such as most Chinese green tea) contain pyrazines, pyrroles and furans, which are formed due to the Maillard reaction. In contrast, the major volatile components in black tea are linalool and its oxides, geraniol, methyl salicylate, (*E*)-2-hexenal, (*Z*)-3-hexenol and (*E*)-2-hexenyl formate.[97] Oolong tea has a high proportion of (Z)-jasmone, jasmine lactone, methyl jasmonate and indole but less proportion of hexenol and hexenol esters, comparatively. Linalool and geraniol are considered as the principal compounds contributing to the flowery odour of all three types of teas.

Lipids

Lipid content in young dry tea shoots ranges from 4% to 9%.[98] The per cent composition of glycolipids in the total lipid is the highest, more than 50%, followed by neutral lipids and phospholipids, which are 35% and 15%, respectively.[99,100] Compounds such as (*Z*)-3-hexenol, its esters and (*E*)-2-hexenal, which primarily contribute to the fresh and greenish odour of green tea, are oxidation products of free fatty acids formed during manufacturing.

3.2 The colour compounds in tea

Chlorophyll

Carotenoids and chlorophyll are the main substrates of the volatile compounds present in processed tea. Total chlorophyll (a and b) content varies from 1 to 7 mg/g in tea shoots.[101] Chlorophyll a and b, and their degradation products, pheophytin a, chlorophyllide and pheophorbide (Table 6), have been found to be present in fresh leaves.[102] Generally, majority of chlorophylls in green tea are preserved during its manufacturing stage, giving the green colour to tea. However, some chlorophylls are transformed to pheophytin and epimers due to the heating treatment during processing. For example, chlorophyll degrades and transforms to either pheophytin or first to chlorophyllide, catalysed by chlorophylase,[103] and finally to pheophorbide. Some chlorophylls remain unchanged, due to which dull infusion and grassy taste are found in black tea.[104–106] Some studies have shown that the influence of chlorophyll on black tea quality might be dependent on levels of other compounds, such as thearubigins and theaflavins.[101] High concentrations of theaflavins can reduce the negative influence of chlorophylls on the appearance of black tea infusions.[104] Oolong tea contains pheophytin b, high concentrations of pyropheophytina and low concentrations of chlorophyll a and b.[107]

Carotenoids

Concentrations of carotenoids in tea shoots range from 0.36 to 0.73 mg/g dry weight. During processing, carotenoid content is reduced via oxidation during withering and

Table 6 Chemical structures of major components of chlorophylls in tea

Mg	R_1	R_2	Chlorophyll species
+	$C_{20}H_{39}$	CH_3COOH	Chlorophyll a
+	H	CH_3COOH	Chlorophyllide a
–	$C_{20}H_{39}$	CH_3COOH	Pheophytin a
–	H	CH_3COOH	Pheophorbide a
–	$C_{20}H_{39}$	H	Pyropheophytin a
–	H	H	Pyropheophorbide a

fermentation or paralytic reaction during firing.[108] Structure of carotenoids is changed through the conversion of violaxanthin to auroxanthin. Volatile compounds of ionine series are formed as primary oxidation products of carotenoids.[108–110]

4 Analytical techniques for tea characterization: overview and chromatic techniques

4.1 Overview

Chemical composition in tea shoots is influenced by their growing surroundings, and its components influence the quality of tea. The next two sections offer an overview on investigation of analytical methods that are being used for the determination of the compounds in tea. Generally, the analytical methods have been divided into three groups: chromatographic techniques, spectroscopic techniques and other techniques (Table 7).

4.2 Chromatographic techniques

During the past decade, the tendency of using chromatographic methods for the determination of compounds in tea has been increasing. These chromatography methods include high-performance thin-layer chromatography (HPTLC), high-performance liquid chromatography (HPLC), ultra high-performance liquid chromatography (UHPLC), capillary electrophoresis (CE), gas chromatography (GC) and their combination with mass spectrometry (MS).

High-performance liquid chromatography (HPLC) and liquid chromatography–mass spectrometry (LC-MS)

Of a wide variety of analytical methods available, HPLC is the most powerful and has shown good separations of the tea phenolic compounds.[111] The HPLC method is widely used in the

Table 7 Overview of analytical methods used in the determination of the compounds in tea

Analytical techniques	Main method	Abbreviation	Main differentiating markers
Chromatography methods	High-performance thin-layer chromatography	HPTLC	Polyphenols(catechin, theaflavin, quercetin and kaempferol triglycerides, and quinic acid esters), amino acids (glutamic acid, asparagine, serine, alanine, leucine, and isoleucine), caffeine, total catechins, Caffey quince acid, kaempferol 3-*O*-rutinoside, kaempferol 3-*O*-acyl glycosides, gallocatechin and quercetin-3-*O*-glucoside
	High-performance liquid chromatography	HPLC	
	Ultra performance liquid chromatography	UPLC	
	Capillary electrophoresis	CE	
	Gas chromatography	GC	
Spectroscopy methods	Nuclear magnetic resonance	NMR	Organic compound; Al, Ba, Ga, Co,Cr, Cs, Gu, Fe, Mg, Mn and Sr
	Near infrared spectroscopy	NIR	
	Atomic absorption spectroscopy	AAS	
	Atomic emission spectroscopy	AES	
Others techniques	Mass spectrometry	MS	Combined with other techniques to detect compounds, like GC–MS; Volatile compounds
	Electronic nose	E-nose	
	Electronic tongue	E-tongue	

determination of numerous compounds, such as tea catechins, gallic acid, purine alkaloids, theanine, among others, in tea because of its high efficiency and high resolution.[112]

HPLC with diode array detection (DAD) is a frequently used method that allows the analysis of polyunsaturated species with high sensitivity and is suitable for the simultaneous separation, identification and quantitation of individual catechins, alkaloids and phenolic compounds in green, oolong, black and Pu-erh teas.[113] However, DAD does not distinguish between different compounds with similar chromophores. In order to obtain more detailed structural information, DAD combined with other techniques, such as MS, or even tandem mass spectrometry (MS/MS), in detecting compounds has become widespread in plant source screening and has also been used in the analysis of tea and its products. For example, an HPLC-DAD-ESI-MS/MS method was applied for analysing young leaves, and sixteen components were identified and 11 major components, including nine catechins and two purine alkaloids, were quantified.[114] However, HPLC coupled with DAD, ESI/MS,

MS/MS and CEAD detection is still the most widely used analytical method for analysing tea compounds.

UHPLC represents another interesting development in separation techniques found in the literature. It is an improved version of the HPLC method. An application of the UHPLC method coupled with time-of-flight mass spectrometer (TOF-MS) was used for the quality assessment of Japanese green tea based on their metabolomic profiling.[115] UHPLC coupled with triple quadruple MS/MS can improve the speed of analysis and provide results with higher sensitivity and accuracy. A multi-residue method based on the application of UHPLC/MS/MS combined with modified QUECHERS (Quick, Easy, Cheap, Effective, Rugged, Safe) sample preparation procedure has been successfully applied for the determination of tea samples. A newly developed UHPLC/DAD/MS method was used for tentative identification and quantification of the major constituents in teas. It has been reported that the UHPLC method has the practical advantages of shorter analysis time and reduced solvent consumption, which makes it an attractive alternative to conventional HPLC technique in fingerprint analysis,[116] especially in situations where high sample throughput and fast analytical speed are required.

Capillary electrophoresis (CE)

CE has played a vital role in various areas of applied analytical chemical and biochemical sciences since its introduction. It was also developed for the rapid analysis of active compounds in green tea samples and Chinese herbs.[117] Among the CE modes, capillary zone electrophoresis (CZE) and micellar electrokinetic chromatography are two widely used modes.

Advantages of CE include: multiple separation modes, such as chromatography-like and electrophoretic elect kinetic separation; minimal need for sample compounds; a broad detection range from UV absorption to MS; and a wide analytical spectrum from inorganic ions to large protein molecules. These advantages mean that Ce is versatile, rugged, economic and rapid, making it an ideal tool for food analysis. Analysis of major amino acids (such as theaflavins in black tea[119], catechins and caffeine in green tea[120]) contained in tea leaves and beverages[118] can be used for the determination of growing areas of green tea leaves based on the catechin profiles.[121] Compared with HPLC, CE not only has the advantage of good separation efficiency and high sensitivity but also requires short analysis time and can process the samples of even small sizes.[122] In conclusion, CE is a useful tool for determining the quality of different kinds of tea samples.

Gas chromatography (GC) and gas chromatography–mass spectrometry (GC–MS)

GC is one of the separation techniques used in tea analysis, mainly in the analysis of volatile and semi-volatile composition, aromas and pesticides. It is possible to classify teas according to their geographical origins by analysing the GC profiling of various compounds present in tea. GC has been the technical choice for analysis of volatile components,[123] oil composition, aromas[124] and pesticides[125] in tea and tea commodities. However, many pesticides which are thermally unstable or non-volatile such as carbamates and benzimidazoles are difficult to analyse with GC. In order to do so, GC combined with other detectors may be a practical method. For example, 19 pesticide residues in tea-growing regions were determined by GC/SIM-MS ; 36 pesticides by GC/time-of-flight-MS (GC/TOF-MS), with the limit of quantitation (LOQ) ranging from 1 to 28 μg/kg[126]; and 85 pesticides, including organochlorine, carbamate, organophosphorus and pyrethroids in vegetables, fruits and green tea are determined by GC/MS.[127]

GC/MS is one of the most important instrumental separation techniques for analysis of volatile and semi-volatile components, such as alcohols, ketones, aldehydes, phenolic, acids, esters among others. The main advantages of GC/MS are its potential to analyse complex mixtures, identify the separated components by using MS, and its high sensitivity and low limit of detection (LOD). For instance, the effects of geographical regions on black tea from the same vegetatively propagated cultivar[128] and the characterization of volatile organic components in black tea were investigated by GC–MS. The ranking of green tea can be predicted by the GC–MS metabolite fingerprinting.[129]

High-performance thin-layer chromatography (HPTLC)

The HPTLC method was applied in the identification of green teas and their extract, for example, for identifying flavonoids in green tea samples from China, Japan and India.[130] It also has protocols for the separation, identification and quantification of polyphenols and polyphenolic acids. A large number of samples can be simultaneously analysed in a short period of time, which is the major advantage of the HPTLC method. The method utilizes lower quantities of solvents, thus lowering the cost of analysis, and the method of detection does not place any restriction on the choice of the mobile phase.

5 Analytical techniques for tea characterization: spectroscopic techniques

Spectroscopic methods, such as nuclear magnetic resonance spectroscopy (NMR), infrared spectroscopy (IR), ultraviolet-visible (UV-Vis) spectroscopy and atomic spectroscopy, such as atomic emission spectroscopy (AES) and atomic absorption spectroscopy (AAS), are equally efficient as chromatographic methods and have been widely used for the quality control of teas.

5.1 Nuclear magnetic resonance spectroscopy (NMR)

NMR is being widely used for food analysis, including the quality evaluation and classification of tea samples. A method for the quality evaluation of green teas according to the metabolite was reported based on NMR spectroscopy.[131] One application is that the quality of Japanese green tea was evaluated by 1H NMR-based global metabolic profiling,[132,133] and the geographical dependencies of grapes and wine metabolome's information, for example, evidence of vintage effect on grape wine, was provided by 1H NMR-based metabolomics.[134–136]

5.2 Infrared spectroscopy (IR)

Infrared spectroscopy (IR) is a fast, accurate, non-destructive analytical tool and also has been applied in the analysis of some active compositions in tea.

Near-infrared spectroscopy (NIR) has been used in the analysis of some active compositions in tea and has major advantages such as high speed, low cost and reliable detection for quantitative and qualitative analyses in various industries, such as agriculture, pharmaceuticals, food, textiles, cosmetics and polymer production. NIR spectroscopy was employed for the analysis of authentic geographical origins of teas;[137] the qualitative identification of tea categories;[138–140] the analysis of the antioxidant capacity of green tea;[141] and simultaneous measurement of the contents of alkaloids and phenolic substances,[142] and caffeine and total polyphenol in green tea.[140] The Vis/NIR method was applied to discriminate tea samples from three different tea gardens[143] and different varieties of instant milk teas.[144] A Fourier transform near-infrared reflectance (FT-NIR) spectroscopy method was developed for the simultaneous analysis of main catechins in green tea.[145] Another application of the FT-NIR method coupled with supervised pattern recognition was reported for the discrimination of Chinese green tea from different places.[145]

5.3 Ultraviolet-Visible (UV-Vis) spectroscopy

Ultraviolet-Visible (UV-Vis) spectroscopy is one of the most common techniques used in routine analysis based on reflectance measurements from 325 to 1125 nm to classify different varieties, which pinpoints the usefulness of the UV-Vis region. Thus, the development of classification methodologies using this instrument is very advantageous due to no additional cost, simplicity or the acquisition of new equipment.

UV-Vis spectroscopy combined with chemometric algorithms has been widely used to differentiate black, green and Pu-erh tea varieties.[146] This method is also applied to quantify the content of caffeine and total polyphenols in tea[147–150] and to realize simultaneous geographical and varietal classification of tea infusions simulating a home-made tea.[151] Another application of UV-Vis spectroscopy combined with machine vision system and artificial neural networks was reported for rapid measurement of concentration of TF and TR in orthodox black tea.[152]

5.4 Atomic emission spectroscopy (AES)/atomic absorption spectroscopy (AAS)

The metal content of teas is influenced by the soil composition and local environmental factors. Several studies including AAS,[153] inductively coupled plasma atomic emission spectrometry (ICP-AES),[154] inductively coupled plasma mass spectrometry (ICP-MS),[155] CE[156] and total reflection X-ray fluorescence spectrometry (TR-XRF) have been carried out in order to determine the levels of minerals in tea leaves.[157] Twelve elements consisting toxic metals (Al, As, Pb, Cr, Cd and Ni) and essential mineral elements (Fe, Zn, Cu, Mn, Ca and Mg) were analysed using inductively coupled plasma atomic emission spectroscopy (ICP-AES).[158] Table 8 gives a brief introduction to separation and detection analysis methods of chemical compounds in tea.

6 Determination of compounds in tea: phenolic compounds and sugars

Various techniques have been employed based on organic constituents and mineral contents present in tea. Analytical approaches have been developed for determining the compounds in combination with and without chemo metrics. The next three sections review analytical methods that are being used for the determination of the compounds in tea, beginning with phenolic compounds and sugars.

6.1 Analysis of phenolic compounds

Conventional methods for the determination of the phenolic compounds in tea rely on the measurement of total phenols, and the browning reactions should be taken into consideration. This measurement is carried out by first extracting them completely without chemical modifications. Different solvent systems have been applied for extraction of polyphenols from plants: boiling water, aqueous mixture of organic solvents such as methanol, acetone, ethanol and N, N *-dimethylformanide*. For example, in ISO 14502, the ground tea is extracted twice with 70% aqueous methanol at 70°C for measurement of total polyphenol and catechins. Boiling water is the most common solvent applied for the extraction of polyphenolics from green, black and maté teas.[122,169,172] Traditional methods for the determination of polyphenols in tea shoots or manufactured teas have been described by several researchers, and the most common methods are paper chromatography, colorimetric measurement and column chromatography. Spectrophotometry is one of the relatively simpler methods for the quantification of phenolic compounds. For example, in ISO 14502-1, the Folin–Ciocalteu methods were the main widely used spectrophotometric assays to measure total phenolic content in plant materials,[181] in which Folin–Ciocalteu reagent is added to the extract after dilution along with a sodium carbonate solution and the blue colour is evaluated using photometry at 765 nm.

As for column chromatography, HPLC is a common technique for both the separation and quantification of phenolic compounds,[182–191] for which sensitivity and detection efficiency are based on purification of phenolic and pre-concentration from complex matrices of crude plant extracts. Generally, a reversed-phase C18 column (RP-C18), photo diode array detector (PDA) and polar acidified organic solvents were used to purify phenolic compounds when detected by HPLC.[192] A small amount of acetic, formic or phosphoric acid is added to the mobile phase to prevent catechins from oxidation and complexion, to adjust the retention time and to improve the peak sharpness of some compounds.

HPLC coupled with MS,[193] electrospray ionization tandem mass spectrometry[194] or CE[122] and CE[120,195] is the most frequently used method to isolate, identify and determine individual polyphenolic compounds, such as (+)-catechin, (–)-epicatechins, theaflavins, alkaloids and caffeine. Research indicates that GC/FID/MS analyses certain types of phenolic compounds with high speed and better resolution than HPLC.

Polyphenols and catechins were also determined by NIR spectroscopy, which showed superiority in speed, efficiency, destructiveness and environmental friendliness.[121]

Table 8 Representation of analytical methods of chemical compounds in tea

Compounds	Separating technology	Detecting technology	Pretreatment (detecting condition, extraction, etc)	Test level	Examples of compounds	References
	TLC	Spectrophotometry	Ninhydrin (pH = 8), 570 nm		Theanine, glutamic acid, glutamine acid, arginine, etc.	44, 115, 159–165
	GC	MS	n-propanol, tallow fatty acid and pentafluorophenol			
	CE	UV/MS	50 nm	10^{-6} mol		
	UHPLC	MS	Postcolumn derivatization, ninhydrin	10^{-12} mol		
	HAEC	PAD				
	HPCEC	Spectrophotometry	Postcolumn derivatization, ninhydrin 440 nm and 570 nm, 56 min	10^{-12} mol		
Amino acids	HPLC	Fluorescence, UV	Dansyl-Cl, 35 min, 6-Aminoquinoly-N-hydroxysuccinimidyl carbamate, 10 min	10^{-13} mol		
		Fluorescence	o-phythalaldehyde, 1 min	10^{-15} mol		
			9-Fluorenylmethyl chlorofornate, 1 min	10^{-15} mol		
		UV	Phenylisothiocyanate, 254 nm, 20 min	10^{-12} mol		
			2,4-Dinitrofluorobenzene, 365 nm, 60 min	10^{-12} mol		
	TLC	Fluorescence	Extraction with water or methanol, GF254, thin-layer chromatogram		Caffeine, theobromine, theophylline, etc.	112, 114, 166–168
	HPLC	UV/MS	C18, 220–280 nm			
		DAD-ESI-MS/MS	Gradient programme from 4 to 25% acetonitrile in 1% formic acid 60 min, 200–700 nm			

Alkaloids	CE	UV	200–210 nm			
	GC	NPD	274 nm			
	TLC, PC	TLC scanner	Extraction with water or methanol		Catechins, flavonoids, anthocyanins, phenolic acids	120, 122, 169–174
	GC	FID/MS	Tetramethylsilane was applied in derivatization			
	UPC2	MS	Extraction with water or methanol			
Phenolics	UHPLC	Q-TOF-MS		High resolution		
	CE	UV	Extraction with water or methanol	High separation efficiency		
	HPLC	Fluorescence detector	SOD, phosphate-borate buffer, 200 nm, 10 min			
		MS or MS/MS	Extraction with water or methanol			
		Chemiluminescence	Extraction with water or methanol			
		Electrochemical	Extraction with water or methanol			
		UV and DAD	Hydrolysis of flavonoid glycosides, 270–285 nm, 35 min			
Chlorophyll and carotenoids	HPLC	UV/DAD	Extraction with dimethylformamide or acetone etc. 440 nm		Chlorophyll, xanthophyll, carotene.	102
	HPTLC	TLC scanner	Acetone ether, acetone, 450 nm			

(Continued)

Table 8 (*Continued*)

Compounds	Separating technology	Detecting technology	Pretreatment (detecting condition, extraction, etc)	Test level	Examples of compounds	References
		Spectrophotometry	Anthrone chemical, enzyme, 470 nm	0.04 μg/mL	Fructose, glucose, sucrose, polysaccharides	175
	HPLC	MS	EIS			
		Refractive index detector	Hypersil NH2 or NH2 columns, or Zorbax carbohydrate	5 mg/g		
		UV	2-aminopyridine, edaravone, 2-anthranilic acid, 245 nm	10^{-12} mol		
Sugar	HPAEC	PAD	0.5 M NaOH and 1.0 M CH_3COONa	μg/L		
	IEC	PAD				
	GC	MS	Alkylation and acetylation			
	HPCE	ELSD	2-Aminopyridine, 9-fluorenylmethyl chloroformate, and edaravone, 200 nm			
	ICP-OES or MS	High-temperature (dry) or concentrated-acid (wet) digestion			Al, Ba, Ga, Co, Cr, Cs, Gu, Fe, Mg, Mn, Sr	176–180
Nutrient elements	AAS	High-temperature (dry) or concentrated-acid (wet) digestion				

*Note: ICP-OES: inductively coupled plasma optical emission spectrometry; HAEC: high-performance anion exchange chromatography; HPCEC: high-performance cation exchange chromatography; UPC2: ultra performance convergence chromatography; HPCE, high-performance CE; AAS, atomic absorption spectrophotometry; IEC, ion exchange chromatography; PAD, pulsed amperometric detector; NPD, nitrogen–phosphorus detection; Q-TOF-MS, quadrupole time-of-flight MS; GPC, gel permeation chromatography.

However, low sensitivity, large relative error and a requisite correction model (calibration) limit its widespread application.

Colorimetric techniques are simple and economical, but they estimate phenolic compound concentrations only above a certain minimum level and do not quantify phenolic compounds individually; however, colorimetric techniques can be valuable for quick and inexpensive screening of numerous samples.[192]

TLC and paper chromatography are two main partitioning techniques applied to separate phenolic compounds in foods. Paper chromatography is a straightforward method and less utilized compared to column chromatography. Paper chromatography has been employed to separate and identify phenolic compounds from tea leaves with butanol/acetic acid/water as the mobile phase. Flavonoids, phenolic acids and glycoflavones have been separated from three green leafy vegetables using paper chromatography. Phenolics can be separated by a series of TLC techniques, which are inexpensive and which can detect multiple phenolics on the same TLC plate in a short analysis time.

As for theaflavins in black tea, isocratic or gradient RP-HPLC[196] is always used for separation. A stabilization solution consisting of EDTA and ascorbic acid is added to the samples to maintain stability of analyses. The formation of oligo-polymeric thearubigins from catechins, theaflavins, catechins and theaflavins and so on has been suggested based on insufficient evidence owing to the complex nature of the multi-component reaction system and multi-reaction sites in each component. There are two ways to isolate TR fractions. One way to isolate TR in black tea fraction is to combine liquid–liquid extraction, for example, countercurrent chromatography (CCC)[197] or high-speed countercurrent chromatography (HSCCC)[198,199] with ethyl acetate/*n*-butanol/water solvent systems. Another approach to isolate TR fraction in precipitation is by adding caffeine[200] to form a TR-caffeine precipitates complex. Recently, matrix-assisted laser desorption/ionization time-of-flight (MALDI-TOF) was used to obtain molecular mass distribution and structural information of TFs and TRs in black tea.[201–203]

6.2 Analysis of sugars

Generally, the anthrone chemical methods can be used as a way of determining the total sugar content. Sugar components such as fructose, glucose and sucrose can be separated by HPLC, high-performance anion exchange chromatography and GC/MS (Table 8). Aminopropyl silica- or potassium-bonded HPLC columns are necessary for the separation of sugars as their chemical and physical properties vary only slightly. The UV absorption efficiency of sugars is low. Therefore, evaporative light-scattering detector (ELSD) and MS are frequently used as detectors.

Qualitative and quantitative determination of polysaccharides in tea is difficult due to their complex structures. TPS is a glycoprotein composed of different monosaccharides, uronic acids and protein, while the glucose or other monosaccharides are usually used as the calibration standard in traditional spectrocolorimetry method that will lead to large deviation from the true value.[204]

There are two methods to determine polysaccharide content in teas. One is an indirect measurement method, based on the condensation reaction by monosaccharide, such as the anthrone–sulphuric acid method or phenol–sulphuric acid method after polysaccharides are transformed to monosaccharides. The second is the direct detection method by HPLC, gel chromatography, density gradient centrifugation and electrophoresis, usually detected by

DAD or electrochemical detector after being separated by capillary zone electrophoresis (CZE). The second method is limited because of difficulties in preparing TPS of high purity because proteins having the same molecular weight as TPS will be mistaken for TPS as they occur in the same peak with TPS. Thus, this method is only adapted to TPS that has relatively higher purity, and to exclude some interfering compounds before assay is a time-consuming process. Generally, only polysaccharides of homogeneous groups within certain molecular weight ranges can be separated and purified.

7 Determination of compounds in tea: analysis of volatile compounds

The quantification of individual flavour compounds is usually performed on GC/MS with flame ionization detection (GC/MS/FID).[205–207] Other methods such as HPLC,[208–211] UHPLC,[212] FT-IR[175,213,214] and electrospray ionization ion mobility spectrometry (ESI-IMS) are also used.[215]

An important procedure prior to instrumental analysis of volatile and semi-volatile compounds is extraction. There are a number of methods to extract volatile compounds, of which the major methods are enumerated as follows: direct organic solvent extraction, simultaneous distillation and extraction, steam distillation under reduced pressure, brewed extraction, adsorptive column method, stir bar adsorptive extraction, solvent-assisted flavour evaporation and solid-phase microextraction.

7.1 Direct organic solvent extraction

Direct organic solvent extraction method has been applied to investigate endogenous volatile compounds in plant tissues.[216] However, non-volatile materials, such as leaf waxes, pigments, lipids and plasticizers from laboratory apparatus, are also likely to be extracted and may complicate analysis or affect the analytical instruments when direct organic solvent extraction method is applied.[217] Hence, selection of extraction solvents, such as hexanes, pentane, diethyl ether, dichloromethane, chloroform, ethyl acetate and solvent mixtures, and use of solid phase extraction column to remove unwanted interfering compounds can be considered in this method.

7.2 Simultaneous distillation and extraction (SDE)

With respect to the extraction of volatile compounds, SDE has been shown to be more advantageous than the other traditional ones: one-step isolation–concentration, adequate reproducibility and high efficiency.[218] The main drawback of SDE is the alteration of some compounds from degradation and hydrolysis or the formation of new products by thermal reactions during the extraction process, which may be due to pollution, oxidation and thermal reactions.[219] Additionally, SDE is a technique applied with a Likens-Nickerson apparatus, which unites the advantages of liquid–liquid and steam distillation–extraction and has been widely used for the extraction of essential oils and volatile compounds from numerous matrices.

7.3 Steam distillation under reduced pressure (SDR)

SDR has been used for the isolation of the volatile compounds including potent odorants from green tea,[220] and aroma concentrate from black tea.[221] Comparing SDR with SDE for the preparation of the aroma concentrates from oolong tea, the aroma constituents from the two methods showed significant differences: the yields of aliphatic aldehydes, aliphatic alcohols, terpene alcohols, esters, α-farnesene and β-ionone of the oolong tea extract prepared by SDE were higher than those prepared by SDR, while the yields of the volatile compounds with relatively high boiling point and heat-sensitive volatile compounds, for example organic acids, lactones, aromatic alcohols and indoles, in the extract prepared by SDR were higher than those prepared by SDE. The drawbacks of SDR are that some glycoside-bound volatile compounds are still hydrolysed and a comparatively long time is required to obtain the tea volatile concentrates.[222]

7.4 Brewed extraction

Brewed extraction is a simple and fast method of sample preparation when compared with SDE. Furthermore, it may be a more appropriate method for reproducing the actual volatile composition; although alterations of aroma compounds cannot be ruled out, they can be minimized using this method. In addition, the yield of concentration of volatile compounds in the brewed extract is less than one-tenth of the yield from SDE. Therefore, it remains to be determined whether brewed extraction can be used to fully extract all the volatile compounds in tea.[222]

7.5 Adsorptive column method

The adsorptive column method is another method for extraction, in which tea sample is extracted by boiled water, and the resulting tea infusion passes through the adsorptive column to isolate the volatile compounds. It has been reported that adsorptive column method was proposed for the extraction of volatile compounds,[18] and that of glycoside-bound volatile compounds from natural green tea.[223,224] The drawback of this method is that large quantities of solvents are needed to fully elute the compounds from the adsorbents. Another drawback relates to the elution capacity and selectivity of the solvents. However, diethyl ether or mixtures of isopentane and diethyl ether were proposed as suitable solvents for extraction.[222]

7.6 Solvent-assisted flavour evaporation (SAFE)

SAFE is a good technique for volatile extraction, and it can be applied to the isolation of volatile compounds from complex matrices.[225] This technique can be combined with the adsorptive column method to remove the non-volatile material and prepare volatile compounds from black tea infusions.[221] High boiling point and some compounds can be recovered at high levels using SAFE. Compared with headspace solid-phase microextraction (HS-SPME), the SAFE resulted in higher recoveries of acid- and hydroxy-containing compounds during the extraction of volatile compounds from apple cider, while the HS-SPME technique resulted in higher recoveries of esters and highly volatile compounds.[226]

7.7 Solid phase microextraction (SPME)

In recent years, SPME, a simple, rapid, solvent-free and low-cost technique, is based on the distribution coefficient of analyses between the sample matrix, the gas phase and the fibre coating with an adsorbent polar or non-polar polymer.[227] SPME has been widely used to analyse the volatile compounds of food, such as wine,[228] bread,[229] fruits[230] and sauce.[231] The effectiveness of SPME is influenced by sampling time, temperature and sample volume. Therefore, maintaining consistency in these factors is crucial for obtaining comparable results. Limitations of SPME sampling are that the samples can be injected only once, and the compounds obtained generally are suitable only for qualitative analysis, but not quantitative analysis.

Applications of SDE and SPME for collecting the volatile compounds from oolong tea were compared, indicating that the SPME technique has higher recoveries of highly volatile compounds such as alkanes.[232] An HS-SPME method was employed for quantifying the aroma compounds, which enabled LOD limits less than 40.8 μg/L and LOQ less than 136.0 μg/L.[233] The other conventional extraction methods applied by different researchers are distillation–liquid/liquid extraction,[218] vacuum hydrodistillation,[234] hydrodistillation[235] and ultrasound-assisted extraction followed by dispersive liquid–liquid microextraction.[123] While in selecting extraction methods, the key objectives are to obtain the sufficient extract and accurately represent the original composition in the sample being analysed. Table 9 summarizes the applications, advantages and drawbacks of various techniques employed for extraction and collection of volatile compounds present in tea.

After the volatile compounds have been collected or extracted, the instruments used for their analysis must be decided. GC/MS is the most efficient technique for the separation, identification and quantification of volatile compounds. Generally, polar or mid-polar GC columns are used to investigate most volatile compounds in tea, while a CHIRAMIX column is used to investigate the stereochemistry of volatile compounds and a multidimensional GC/GC/MS is applied to separate target components from complex mixtures. For example, GC/GC/MS system was used to separate (*R*)- and (*S*)-1-phenylethanol in a ratio of 96:4 in the endogenous volatiles in leaves of *C. sinensis* var. Yabukita.[236] GC coupled with olfactometry methods is effective in determining the odour-active compounds and their contribution to the aroma quality, and in evaluating the flavour changes that occur during the manufacturing of tea beverages.[220] Currently, aroma extraction dilution analysis (AEDA) is most frequently used for evaluating odour-active compounds in tea analysis. For example, 4-methoxy-2-methyl-2-butanethiol (meaty), (*Z*)-1,5-octadien-3-one (metallic), 4-mercapto-4-methyl-2-pentanone (meaty), (*E*, *E*)-2,4-decadienal (fatty), β-damascone (honey-like), β-damascenone (honey-like), (*Z*)-methyl jasmonate (floral) and indole (animal-like) were identified as the most important components of the common Japanese green tea (Sen-cha) odour[220]; 2-acetyl-1-pyrroline (popcorn-like), 2-ethyl-3,5-dimethylpyrazine (nutty), 2,3-diethyl-5-methylpyrazine (nutty) and 2-acetyl-2-thiazoline (popcorn-like) were detected as the most potent odorants only in the pan-fired green teas,[239] and (*E*, *E*)-Nona-2,4-dienal and α-pinene were characterized as additional odorants of black tea.[240] Generally, all peaks are identified by comparing them with those of standard compounds, or by searching the commercial NIST library, and use of retention indices. For relative quantification, ethyl n-decanoate is often used as an internal standard for calculating the relative peak area ratios of the target volatile compounds and the internal standard.

Table 9 Summary of techniques for tea endogenous volatile compound extraction and collection

Extraction	Application	Advantages	Drawbacks	Reference
Direct organic solvent extraction	Analysis of volatile compounds in fresh tea leaves	Enzymatic alteration of volatile extraction is minimized	Non-violation compounds such as caffeine are extracted and subjected to GC–MS affecting the GC-column	236
Simultaneous distillation and extraction (SDE)	Analysis of volatile compounds in tea leaves without enzyme activity during the manufacturing process and in tea products	Water extraction and organic solvent extraction are simultaneous. The yield of volatiles can be increased	Due to the high extraction temperature, the volatile components are altered, for example, some precursors are hydrolysed	222
Steam distillation under reduced pressure (SDR)	Analysis of volatile compounds in tea products and tea leaves without enzyme activity during the manufacturing process	A simple method for obtaining large yields of essential oils. The aroma characteristic of the original tea sample is relatively well reflected	Some glycosidically bound volatile compounds are still hydrolysed and a comparatively long time to obtain the tea volatile concentrates is required	220
Brewed extraction	Analysis of volatile compounds in tea products and tea leaves without enzyme activity during the manufacturing process	Simple and fast	Due to the high extraction temperature, hydrolysis of precursors such as glycosides cannot be avoided	222, 237
Adsorptive column method	Analysis of volatile compounds in water extracts of tea or tea beverages	Some non-volatile compounds can be removed prior to analysis by GC–MS	Different organic solvents used for elution may result in alteration of the volatile pattern	222
Solvent-assisted flavour evaporation (SAFE)	Analysis of volatile compounds in water extracts of tea or tea beverages	Faster and more sensitive than conventional techniques	A thermal desorption unit is required	222, 237
Solid phase microextraction (SPME)	Collect volatile compounds from dry tea samples and tea infusion; investigate the volatile compounds emitted from leaves of tea plants	Fast, effective and simple method. Non-polar semi-volatile and volatile compounds can be effectively extracted with non-polar fiber coatings such as PDMS	The compounds obtained are generally fit for qualitative analysis, but not for quantitative analysis. The yields are sufficient for GC analysis, but not for the structural elucidation of unknown compounds	238

Recently, a faster analytic method, E-nose, which consists of a sampling apparatus, a trap tube, a detector unit containing the array of sensors, a computer system for data recording and analysis and transportable z-Nose, which is an ultrafast portable GC-based volatile compound analyser with high sensitivity, has been used to analyse volatile compounds in tea. For example, E-nose has been widely used to predict tea quality,[241] monitor the black tea fermentation process,[242] identify tea grades[243] and evaluate particular tea flavours.[244]

8 Determination of compounds in tea: other compounds and elements

8.1 Analysis of chlorophyll and carotenoids

Spectrophotometry is often used in the determination of chlorophyll content because it absorbs different spectra at different wavelengths. HPLC with a PDA detector was used for the separation and identification of pigments, chlorophylls and carotenoids of seven teas, and 79 pigment species, and 41 chlorophylls and 38 carotenoids were detected.[102] *N, N*-Dimethylformamide and acetone are usually used as solvents for pigment extraction, while the DAD detector is a good choice for identifying the spectral characteristics of chlorophyll and carotenoids.

8.2 Analysis of nitrogen-containing compounds

Amino acids

There are two methods for detecting amino acids present in tea. The first method is a direct analysis with HPLC, CE or MS/MS analysis. An example is the ISO 19563:2017 standard for determining theanine in tea and instant tea in solid form using HPLC. The second method is a chemical derivatization approach. For example, the precolumn derivatization technology gained considerable acceptance. This technique utilizes AccQ•tag (6-aminoquinolyl-*N*-hydroxysuccinimidyl) carbamate to transform primary and secondary amines into highly stable fluorescent derivatives in less than 10 min, which could be reserved at 25°C for seven days.[159,245,246] This technology can also be used in conjunction with MS or UHPLC. Comparing the performance of PicoTag-HPLC with that of AccQ•Tag-UHPLC in evaluating amino acids, the AccQ•Tag-UHPLC method improved LOD up to three times and reduced the analysis time by 2.5 folds.[159] AccQ•Tag-UPLC-ESI-MS/MS, applied to amino acid detection and quantitation, is a fast and sensitive method.[247]

In heating–derivatization conditions, amino acids produce purple complex with ninhydrin (pH = 8.0). Then, the total content of amino acids will be determined by a spectrophotometer at 570 nm. The amino acids are quantitatively and qualitatively analysed after pre-separation with paper chromatography, TLC or high-performance cation exchange chromatography. The advantages of these methods are their short separation time and high column efficiency, while the disadvantage is the fact that the derivative reaction is susceptible to interference.

Alkaloids

The alkaloids can be simply and rapidly determined by spectrophotometry, for example, in ISO 10727-2002-07-15, the determination of the caffeine content of teas and instant teas can be made by HPLC. HPLC coupled with a UV or MS detector is a popularly adopted technique (Table 8). Another method uses a reversed-phase HPLC system for the separation of purine alkaloids: caffeine, theophylline and theobromine.[112] It has been reported that HPLC-MS can be used to test for caffeine, theobromine and theophylline.[167] The LOQ of caffeine, theobromine and theophylline determined by HPLC coupled with a MS detector is less than 0.1 ng.

Attempts to simultaneously measure water, alkaloids and phenolic substance content in tea leaves have been made using NIR spectroscopy.[142,248] An FT-NIR spectroscopy method was developed for the detection of caffeine content in instant green tea powder.[249] In HPLC analysis, under the condition of gradient elution (acetonitrile and water), the alkaloids can be separated and detected synchronously with catechins in a C18 column.[250] Caffeine can also be extracted, pre-concentrated and analysed by dispersive liquid–liquid microextraction and gas chromatography–nitrogen phosphorus detection, which is a simple, fast and green method. The LOD is 20 ng/mL and LOQ is 50 ng/mL.[168]

8.3 Analysis of tea saponins

The early determination of saponins present in plant material was based predominantly on gravimetry or on methods taking advantage of some of their chemical or biological features. However, saponins do not have any characteristic functional group that can react with a colouring agent; hence it is difficult to determine their amount by colorimetry. Quantitative determination of saponin in a tea plant has also been difficult without using their purified standards, because each plant contains more than dozens of saponins, making it difficult to obtain all the standards and to estimate the total amount of saponins in tea.

TSs have identified theasapogenol B as an aglycone, and most of them contain a cinnamoyl group as a functional group.[90–92] It has been analysed and detected by different methods. For example, saponins were determined by UPLC/Q-TOF-MS, LC/TOF-MS and quantified by LC/UV.[251,252] Matsui considered that the fragment peak of theasapogenol B measured by LC/TOF-MS could be used to identify the saponins and that the absorption at 280 nm of the cinnamoyl group measured by LC/UV could be used to quantify the saponins.[251] A rapid pressurized liquid extraction and HPLC-DAD–MS method with a Prevail C18 rocket column (33 mm × 7 mm, 3.0 m) and gradient elution was developed for the simultaneous determination of flavonoid, saponins and polyacetylenes in Ginseng.[253]

8.4 Analysis of elements in tea

Tea contains a variety of elements, such as nitrogen, carbon, sulphur, hydrogen and metals. Table 10 shows different elements present in made tea and tea infusions that are categorized into major, minor and trace elements. In general, atomic and MS are used for the analysis of elements in tea. These include flame-and-graphite furnace atomic absorption spectrometry (FAAS)[179] and graphite furnace atomic absorption

Table 10 Elemental contents of made teas[258]

	Elements	Made tea (μg/g)	Tea infusions (μg/g)
Major	Al	(0.02–27.0)*10^3	2.6–63.3
	Ca	(n.d–4.60)*10^4	nd–23.2
	K	(0.63–2.99)*10^4	0.2–100
	Mg	(0.09–34.0)*10^3	1.5–54.6
	Mn	(nd–1.50)*10^3	0.9–73.9
	Na	3.00–3.20*10^4	0.6–96.9
	P	4.84–5.79*10^3	nd–54.4
	S	(2.04–3.83)*10^3	23.1–35.0
Minor	B	3.10–57.8	0.1–100
	Ba	0.57–63.0	nd–7.6
	Cr	n.d-129	nd–67.5
	Cu	0.03–602	2.7–74.8
	Fe	0.03–1.30*10^4	0.1–57.0
	Mo	nd–571	14.6
	Ni	nd–1.96*10^3	0.1–100
	Rb	23.1–152	41.9–100
	Sr	0.16–72.4	0.5–30.9
	Ti	nd–263	<3
	Zn	nd–1.12*10^3	0.1–88.8
Trace	Ag	nd–0.05	nd
	As	nd–10.2	nd–44.9
	Be	nd	nd
	Bi	nd–0.28	nd–1.2
	Cd	nd–8.06	nd–100
	Ce	0.51–0.59	–
	Co	nd–17.4	0.9–69.8
	Cs	0.08–0.84	–
	Ca	nd–2.03	–
	Hg	nd–0.76	nd
	In	nd–0.92	nd–0.30
	Li	nd–16.0	nd–0.20

Pb	nd–240	nd–58.6
REEs	nd–4.22	nd–12.9
Sb	nd–0.08	0.3–18.9
Se	nd–16.4	nd–52.5
Sn	nd–0.81	nd–2.2
Te	–	nd–0.16
Tl	0.03–10.7	nd–79.4
U	–	0.006
W	–	0.003
V	nd–29.0	nd–18.7
Y	0.36	11.9
Zr	nd–0.70	4.9

Note: nd: not detected; REEs: rare earth elements.

spectrometry[254] inductively coupled plasma optical emission spectrometry (ICP-OES)[177–180,253–255] and ICP-MS,[256] neutron activation analysis.[257] The ICP-based techniques are the most effective in the multi-element analysis of tea samples. In the case of ICP-MS, information on isotopic distribution is obtained. Hydride generation combined with a spectrometric detector, that is, atomic fluorescence spectrometry, is preferred for the determination of hydride-forming elements, such as As, Bi, Se and Te. For other elements, the total nitrogen is commonly determined by the Kjeldahl method, and the Dumas combustion method on an elemental analyser was used for the analyses of nitrogen, carbon, sulphur and hydrogen.

9 Diversified tea products

History bears witness to tea incarnations from loose tea to blend tea, tea bag, tea packet, instant tea, ready-to-drink (RTD) tea, and flavoured teas,[259] which all are popular products on the market.

9.1 Instant tea

Instant tea, a new-fashioned fast-growing consumer product in every country, is a kind of solid beverage tea that can dissolve quickly in water without residual of tea powder, granular, or flake. Besides that, instant tea has various advantages, for example, small volume, minimal pesticide residues and good taste. It is made using raw materials, such as the finished product tea, semi-finished tea, tea by-product or fresh leaf. Generally, instant tea is divided into two categories: pure tea and spices tea; pure tea is made up of instant black tea, instant green tea, instant oolong tea, instant jasmine tea or instant Pu-erh tea, while spices tea consists of sugar-containing black tea, green tea, oolong tea and lemon black tea, milk tea or some kind of fruit instant tea.

Spray-, freeze- and vacuum-drying are the three conventional methods for producing instant tea powders. Spray-drying is a well-established technique to produce powder through liquid and semiliquid food as materials,[260,261] hot air is adopted during concurrent spray-drying to remove a small amount of liquid droplets. During this process, some properties such as colour, flavour and nutrients are retained, because it is a short and controllable process. For example, modified starches, maltodextrin and arabic gum have been applied to the spray-drying process for providing stability and improving product recovery.[262] Freeze-drying, consisting of removing water from a frozen sample by sublimation and desorption under vacuum, is an industrial process. Most of the existing methods for the production of instant tea use hot water extraction. The taste, aroma and strength of tea are determined by the retention of soluble components. The existing extraction method conducted at higher temperatures results in a substantial loss of tea aroma and brings undesirable alterations to the characteristic flavour and appearance of tea powder. For example, volatile components, such as tea polyphenols, are thermally unstable and may undergo degradation during thermal extraction, which directly influences the flavour, taste and colour of tea.[263] Table 11 shows the main compounds and contents present in an instant black tea.

9.2 Ready-to-drink tea

The RTD tea market has become the most dynamic category in the soft drinks industry worldwide. RTD tea can be prepared by tea leaves which undergo extraction, filtration, clarification process, etc... In addition, RTD tea also can be obtained by commingling tea brew with water, sugar, sour agent, edible flavor, fruit juice and vegetable products. It inherits the unique flavour of tea, as it contains natural tea polyphenols, caffeine and other active ingredients, and has both nutritional and health benefits. Thus, it becomes a favoured multifunctional beverage. L-theanine content in RTD tea can be a reliable parameter for evaluating its quality,[212] since L-theanine was reported to have a complex taste profile and have an influence on the basic tastes such as sweet, bitter and umami.[265,266] Table 12 shows the main compounds and other contents present in RTD tea.

10 Summary

The chemical compositions in tea such as flavanols, sugars, alkaloids, amino acids, minerals and phenolic compounds determine the health functions and sensory quality of tea. With the development and improvement of separation and detection technologies and chemometric methods, the qualification and quantitation methods of compounds in tea have become more sensitive, reliable, fast and readily accessible, especially when combined with different analytical techniques or methods. These combined techniques might be significantly more useful than relying on signal analysis technique, because all these technologies have their own advantages and disadvantages in their applications for tea analysis. In addition, tea also appears in a variety of incarnations, which are differently favoured by different populations and analysis methods. This chapter concludes that it is critically important that the properties of tea are closely associated with their profiles of phytochemicals. Therefore, the analysis and profiling of compounds in tea are vital in addressing the biological and flavour properties of a variety of tea.

Table 11 Main contents in an instant black tea[264]

	Compounds	Contents
Catechins	Catechin	288–316 mg/100g
	Epicatechin	524–889 mg/100g
	Epicatechin gallate	315–428 mg/100g
	Epigallocatechin	1004–3460 mg/100g
	Epigallocatechin gallate	173–176 mg/100g
	Gallocatechin	397–506 mg/100g
Theaflavins	Theaflavin	56–124 mg/100g
	Theaflavin-3,3′-digallate	24–27 mg/100g
	Thearubigins	10161–12701 mg/100g
Phenolic acids	Gallic acids	697–751 mg/100g
Alkaloids	Caffeine	3964–4398 mg/100g
	Theabromine	37–44 mg/100g
Sugars	Fructose	6.92–7.62 g/100g
	Glucose	3.30–4.14 g/100g
	Sucrose	0.95–2.48 g/100g
	Total	12.7 g/100g
Amino acids	Alanine	92.7–105.1 mg/100g
	Arginine	231.9–255.4 mg/100g
	Asparagine	76.3–78.6 mg/100g
	Aspartic acid	181.8–209.7 mg/100g
	Glutamic acid	219.9–241.1 mg/100g
	Glutamine	56.6–97.2 mg/100g
	Glycine	1.6–4.0 mg/100g
	Histidine	14.5–14.8 mg/100g
	Isoleucine	29.0–49.7 mg/100g
	Leucine	41.3–63.3 mg/100g
	Lysine	21.1–32.2 mg/100g
	Phenylalanine	29.0–69.8 mg/100g
	Proline	27.6–46.0 mg/100g
	Serine	87.1–109.3 mg/100g
	Theanine	512.9–572.2 mg/100g
	Tryptophan	34.0–54.7 mg/100g
	Tyrosine	10.2–17.5 mg/100g
	Valine	15.4–25.1 mg/100g

(Continued)

Table 11 (*Continued*)

	Compounds	Contents
	Total	1719–2010 mg/100g
Organic acids	Citric	0.83–1.18 g/100g
	Fumaric	0.01 g/100g
	Malic	2.07–2.32 g/100g
	Oxalic	1.79–2.00 g/100g
	Tannic	20.43–21.89 g/100g
	Total	25.58–26.94 g/100g
Composition	Moisture	8.54–9.26 g/100g
	Ash	9.57–10.77 g/100g
	Protein	19.31–19.86 g/100g
	Dietary fibre	1.28–3.05 g/100g
	Dietary soluble	0.19–1.81 g/100g
	Dietary insoluble	1.09–1.24 g/100g
	Fat	0.74–1.10 g/100g
	Carbohydrate	56.68–59.84 g/100g
Minerals	Calcium	73.0–103 mg/100g
	Chromium	0.16–0.17 mg/100g
	Copper	0.47–0.99 mg/100g
	Cobalt	0.06–0.07 mg/100g
	Iron	2.46–3.22 mg/100g
	Magnesium	179–280 mg/100g
	Manganese	151–166 mg/100g
	Molybdenum	0.21–0.42 mg/100g
	Phosphorus	376–431 mg/100g
	Potassium	3928–4625 mg/100g
	Selenium	Trace
	Sodium	4.41–5.41 mg/100g
	Zinc	1.78–2.85 mg/100g
Vitamins	Thiamin	0.09–0.13 mg/100g
	Niacin	20.64–20.85 mg/100g
	Pantothenic	59.08–62.38 mg/100g
	Pyridoxine	0.37–0.46 mg/100g
Carotenoids	Lutein	0.51–1.24 mg/100g

Table 12 Compounds and contents of ready-to-drink green tea[267–270]

Compounds	Contents (mg/L)	Threshold (mg/L)
Gallic acid	2.53–16.0	34
ECGC	16.7–85.4	87
GCG	11.5–68.7	179
EGC	9.96–34.6	159
GC	19.1–56.2	165
ECG	4.77–8.36	115
CG	4.75–31.8	111
EC	7.55–28.2	270
C	6077–24.3	11
Total catechins	84.1–355	–
Caffeine	97.1–243	97
Asp	13.3–14.5	24
Ser	2.26–8.35	1500
Glu	8.65–22.3	9
Gly	0–1.18	1300
His	0.590–3.68	191
Arg	3.36–11.63	209
Thr	2.44–1.23	2600
Ala	0.946–4.58	600
Pro	0.409–1.63	1738
Thea	19.7–42.1	1045
Cys	n.d.	–
Tyr	1.07–3.47	906
Val	1.27–3.41	400
Met	n.d.	–
Lys	0.42–5.35	104
Ile	0.60–1.60	900
Leu	0.424–1.01	846
Phe	0–2.70	900
Total amino acids	45.6–99.4	–
GMP	0–1.10	109
IMP	n.d.	–

11 References

1. Butt, M. S. and Sultan, M. T. 2009. Green tea: Nature's defense against malignancies. *Crit. Rev. Food Sci. Nutr.* 49(5):463–73.
2. Dias, T., Tomás, G., Teixeira, N., Alves, M., Oliveira, P. and Silva, B. 2013. White tea (*Camellia sinensis* (L.)): Antioxidant properties and beneficial health effects. *Int. J. Food Sci. Nutr. Diet.* 2(2):19–26.
3. Rusak, G., Komes, D., Likić, S., Horžić, D. and Kovač, M. 2008. Phenolic content and antioxidative capacity of green and white tea extracts depending on extraction conditions and the solvent used. *Food Chem.* 110(4):852–8.
4. Alcázar, A., Ballesteros, O., Jurado, J. M., et al. 2007. Differentiation of green, white, black, Oolong, and Pu-erh teas according to their free amino acids content. *J. Agric. Food Chem.* 55(15):5960–65.
5. Venditti, E., Bacchetti, T., Tiano, L., Carloni, P., Greci, L. and Damiani, E. 2010. Hot vs. cold water steeping of different teas: Do they affect antioxidant activity? *Food Chem.* 119(4):1597–604.
6. Hilal, Y. and Engelhardt, U. 2007. Characterisation of white tea – Comparison to green and black tea. *J. für Verbraucherschutz und Lebensmittelsicherheit.* 2(4):414–21.
7. Dias, T. R., Alves, M. G., Tomas, G. D., Socorro, S., Silva, B. M. and Oliveira, P. F. 2014. White tea as a promising antioxidant medium additive for sperm storage at room temperature: A comparative study with green tea. *J. Agric. Food Chem.* 62(3):608–17.
8. Komes, D., Horzic, D., Belscak, A., Kovacevic Ganic, K. and Bljak A. 2009. Determination of caffeine content in tea and maté tea by using different methods. *Czech J. Food Sci.* 27:S213–16.
9. Gondoin, A., Grussu, D., Stewart, D. and McDougall, G. J. 2010. White and green tea polyphenols inhibit pancreatic lipase *in vitro*. *Food Res. Int.* 43(5):1537–44.
10. Santana-Rios, G., Orner, G. A., Amantana, A., Provost, C., Wu, S.-Y. and Dashwood, R. H. 2001. Potent antimutagenic activity of white tea in comparison with green tea in the Salmonella assay. *Mutat. Res.* 495(1–2):61–74.
11. Santana-Rios, G., Orner, G. A., Xu, M., Izquierdo-Pulido, M. and Dashwood, R. H. 2001. Inhibition by white tea of 2-amino-1-methyl-6-phenylimidazo [4, 5-b] pyridine-induced colonic aberrant crypts in the F344 rat. *Nutr. Cancer* 41(1–2):98–103.
12. Orner, G. A., Dashwood, W.-M., Blum, C. A., et al. 2002. Response of Apc min and A33 ΔNβ-cat mutant mice to treatment with tea, sulindac, and 2-amino-1-methyl-6-phenylimidazo [4, 5-b] pyridine (PhIP). *Mutat. Res.* 506:121–7.
13. Dashwood, W.-M., Orner, G. A. and Dashwood, R. H. 2002. Inhibition of β-catenin/Tcf activity by white tea, green tea, and epigallocatechin-3-gallate (EGCG): Minor contribution of H_2O_2 at physiologically relevant EGCG concentrations. *Biochem. Biophys. Res. Commun.* 296(3):584–8.
14. Dias, T. R., Alves, M. G., Tomás, G. D., Socorro, S., Silva, B. M. and Oliveira, P. F. 2014. White tea as a promising antioxidant medium additive for sperm storage at room temperature: A comparative study with green tea. *J. Agric. Food Chem.* 62(3):608–17.
15. Moderno, P. M., Carvalho, M. and Silva, B. M. 2009. Recent patents on *Camellia sinensis*: Source of health promoting compounds. *Pat. Food Nutr. Agric.* 1(3):182–92.
16. Zhang, Q. and Ruan, J. 2016. Tea: Analysis and tasting. In: Caballero, B., Finglas, P. and Toldra, F. (Eds), *Encyclopedia of Food & Health*, Academic Press, Oxford, UK, pp. 256–67.
17. Kuo, P. C., Lai, Y. Y., Chen, Y. J., Yang, W. H. and Tzen, J. T. 2011. Changes in volatile compounds upon aging and drying in oolong tea production. *J. Sci. Food Agric.* 91(2):293–301.
18. Sai, V., Chaturvedula, P. and Prakash, I. 2011. The aroma, taste, color and bioactive constituents of tea. *J. Med. Plants Res.* 5(11):2110–24.
19. Lin, C.-C., Li, C.-W., Shih, Y.-T. and Chuang, L.-T. 2014. Antioxidant and anti-inflammatory properties of lower-polymerized polyphenols in oolong tea. *Int. J. Food Prop.* 17(4):752–64.
20. Chen, C.-N., Liang, C.-M., Lai, J.-R., Tsai, Y.-J., Tsay, J.-S. and Lin, J.-K. 2003. Capillary electrophoretic determination of theanine, caffeine, and catechins in fresh tea leaves and oolong tea and their effects on rat neurosphere adhesion and migration. *J. Agric. Food Chem.* 51(25):7495–503.

21. Katiyar, S. K. and Mukhtar, K. 1996. Tea in chemoprevention of cancer: Epidemiologic and experimental studies (Review). *Int. J. Oncol.* 8:221–38.
22. Alasalvar, C., Topal, B., Serpen, A., Bahar, B., Pelvan, E. and Gökmen, V. 2012. Flavor characteristics of seven grades of Black Tea produced in Turkey. *J. Agric. Food Chem.* 60(25):6323.
23. Sarkar, A. and Bhaduri, A. 2001. Black tea is a powerful chemopreventor of reactive oxygen and nitrogen species: Comparison with its individual catechin constituents and green tea. *Biochem. Biophys. Res. Comm.* 284(1):173–8.
24. Aneja, R., Odoms, K., Denenberg, A. G. and Wong, H. R. 2004. Theaflavin, a black tea extract, is a novel anti-inflammatory compound. *Crit. Care Med.* 32(10):2097–103.
25. Pan, M.-H., Lin-Shiau, S.-Y., Ho, C.-T., Lin, J.-H. and Lin, J.-K. 2000. Suppression of lipopolysaccharide-induced nuclear factor-κB activity by theaflavin-3, 3′-digallate from black tea and other polyphenols through down-regulation of IκB kinase activity in macrophages. *Biochem. Pharmacol.* 59(4):357–67.
26. Huang, M. T., Liu, Y., Ramji, D., et al. 2006. Inhibitory effects of black tea theaflavin derivatives on 12-O-tetradecanoylphorbol-13-acetate-induced inflammation and arachidonic acid metabolism in mouse ears. *Mol. Nutr. Food Res.* 50(2):115–22.
27. Song, Y.-A., Park, Y.-L., Yoon, S.-H., et al. 2011. Black tea polyphenol theaflavin suppresses LPS-induced ICAM-1 and VCAM-1 expression via blockage of NF-κB and JNK activation in intestinal epithelial cells. *Inflamm. Res.* 60(5):493–500.
28. Ukil, A., Maity, S. and Das, P. K. 2006. Protection from experimental colitis by theaflavin-3, 3′-digallate correlates with inhibition of IKK and NF-κB activation. *Br. J. Pharmacol.* 149(1):121–31.
29. Yoshino, K., Yamazaki, K. and Sano, M. 2010. Preventive effects of black tea theaflavins against mouse type IV allergy. *J. Sci. Food Agric.* 90(12):1983–7.
30. Hosokawa, Y., Hosokawa, I., Shindo, S., et al. 2011. Black tea polyphenol inhibits CXCL10 production in oncostatin M-stimulated human gingival fibroblasts. *Int. Immunopharmacol.* 11(6):670–4.
31. Hosokawa, Y., Hosokawa, I., Ozaki, K., Nakanishi, T., Nakae, H. and Matsuo, T. 2010. Tea polyphenols inhibit IL-6 production in tumor necrosis factor superfamily 14-stimulated human gingival fibroblasts. *Mol. Nutr. Food Res.* 54(Suppl. 2):S151–8.
32. Das, A. S., Mukherjee, M., Das, D. and Mitra, C. 2009. Protective action of aqueous black tea (*Camellia sinensis*) extract (BTE) against ovariectomy-induced oxidative stress of mononuclear cells and its associated progression of bone loss. *Phytother Res.* 23(9):1287–94.
33. Karmakar, S., Majumdar, S., Maiti, A., et al. 2011. Protective role of black tea extract against nonalcoholic steatohepatitis-induced skeletal dysfunction. *J. Osteoporosis.* 2011(426863), 1–12.
34. Oka, Y., Iwai, S., Amano, H., et al. 2012. Tea polyphenols inhibit rat osteoclast formation and differentiation. *J. Pharmacol Sci.* 118(1):55–64.
35. Kapiszewska, M., Miskiewicz, M., Ellison, P. T., Thune, I. and Jasienska, G. 2006. High tea consumption diminishes salivary 17β-estradiol concentration in Polish women. *Br. J. Nutr.* 95(5):989–95.
36. Saha, P. and Das, S. 2002. Elimination of deleterious effects of free radicals in murine skin carcinogenesis by black tea infusion, theaflavins & epigallocatechin gallate. *Asian Pac. J. Cancer Prev.* 3(3):225–30.
37. Roy, P., Nigam, N., George, J., Srivastava, S. and Shukla, Y. 2009. Induction of apoptosis by tea polyphenols mediated through mitochondrial cell death pathway in mouse skin tumors. *Cancer Biol Ther.* 8(13):1281–7.
38. Patel, R., Ingle, A. and Maru, G. B. 2008. Polymeric black tea polyphenols inhibit 1, 2-dimethylhydrazine induced colorectal carcinogenesis by inhibiting cell proliferation via Wnt/β-catenin pathway. *Toxicol. Appl. Pharmacol.* 227(1):136–46.
39. Mohan, K. C., Hara, Y., Abraham, S. and Nagini, S. 2005. Comparative evaluation of the chemopreventive efficacy of green and black tea polyphenols in the hamster buccal pouch carcinogenesis model. *Clin. Biochem.* 38(10):879–86.
40. Bhattacharya, U., Halder, B., Mukhopadhyay, S. and Giri, A. K. 2009. Role of oxidation-triggered activation of JNK and p38 MAPK in black tea polyphenols induced apoptotic death of A375 cells. *Cancer Sci.* 100(10):1971–8.

41. Sil, H., Sen, T., Moulik, S. and Chatterjee, A. 2010. Black tea polyphenol (theaflavin) downregulates MMP-2 in human melanoma cell line A375 by involving multiple regulatory molecules. *J. Environ. Pathol. Toxicol. Oncol.* 29(1):55–68.
42. Lahiry, L., Saha, B., Chakraborty, J., et al. 2009. Theaflavins target Fas/caspase-8 and Akt/pBad pathways to induce apoptosis in p53-mutated human breast cancer cells. *Carcinogenesis* 31(2):259–68.
43. Yeh, C., Chen, W., Chiang, C., Lin-Shiau, S. and Lin, J. 2003. Suppression of fatty acid synthase in MCF-7 breast cancer cells by tea and tea polyphenols: A possible mechanism for their hypolipidemic effects. *Pharmacogenomics J.* 3(5):267.
44. Banerjee, S., Manna, S., Mukherjee, S., Pal, D., Panda, C. K. and Das S. 2006. Black tea polyphenols restrict benzopyrene-induced mouse lung cancer progression through inhibition of Cox-2 and induction of caspase-3 expression. *Asian Pac. J. Cancer Prev.* 7(4):661–6.
45. Banerjee, S., Manna, S., Saha, P., Panda, C. K. and Das, S. 2005. Black tea polyphenols suppress cell proliferation and induce apoptosis during benzo (a) pyrene-induced lung carcinogenesis. *Eur. J. Cancer Prev.* 14(3):215–21.
46. Yang, G., Liu, Z., Seril, D. N., et al. 1997. Black tea constituents, theaflavins, inhibit 4-(methylnitrosamino)-1-(3-pyridyl)-1-butanone (NNK)-induced lung tumorigenesis in A/J mice. *Carcinogenesis.* 18(12):2361–5.
47. Gosslau, A., Jao, E., Li, D., et al. 2011. Effects of the black tea polyphenol theaflavin-2 on apoptotic and inflammatory pathways *in vitro* and *in vivo*. *Mol. Nutr. Food Res.* 55(2):198–208.
48. Lu, J., Ho, C.-T., Ghai, G. and Chen, K. Y. 2000. Differential effects of theaflavin monogallates on cell growth, apoptosis, and Cox-2 gene expression in cancerous versus normal cells. *Cancer Res.* 60(22):6465–71.
49. Pan, M.-H., Liang, Y.-C., Lin-Shiau, S.-Y., Zhu, N.-Q., Ho, C.-T., Lin, J.-K. 2000. Induction of apoptosis by the oolong tea polyphenol theasinensin A through cytochrome c release and activation of caspase-9 and caspase-3 in human U937 cells. *J. Agric. Food Chem.* 48(12):6337–46.
50. Liang, Y.-C., Chen, Y.-C., Lin, Y.-L., Lin-Shiau, S.-Y., Ho, C.-T. and Lin, J.-K. 1999. Suppression of extracellular signals and cell proliferation by the black tea polyphenol, theaflavin-3, 3′-digallate. *Carcinogenesis* 20(4):733–6.
51. Mizuno, H., Cho, Y. Y., Zhu, F., et al. 2006. Theaflavin-3, 3′-digallate induces epidermal growth factor receptor downregulation. *Mol Carcinog.* 45(3):204–12.
52. Schuck, A. G., Ausubel, M. B., Zuckerbraun, H. L. and Babich, H. 2008. Theaflavin-3, 3′-digallate, a component of black tea: An inducer of oxidative stress and apoptosis. *Toxicol In Vitro.* 22(3):598–609.
53. Li, W., Wu, J.-X. and Tu, Y.-Y. 2010. Synergistic effects of tea polyphenols and ascorbic acid on human lung adenocarcinoma SPC-A-1 cells. *J. Zhejiang Univ. Sci. B* 11(6):458–64.
54. Arts, I. C., Jacobs Jr., D. R., Harnack, L. J., Gross, M. and Folsom, A. R. 2001. Dietary catechins in relation to coronary heart disease death among postmenopausal women. *Epidemiology* 12(6):668–75.
55. Arts, I. C., Hollman, P. C., Feskens, E. J., De Mesquita, H. B. B. and Kromhout, D. 2001. Catechin intake might explain the inverse relation between tea consumption and ischemic heart disease: The Zutphen Elderly Study. *Am. J. Clin. Nutr.* 74(2):227–32.
56. Leung, L. K., Su, Y., Chen, R., Zhang, Z., Huang, Y. and Chen, Z.-Y. 2001. Theaflavins in black tea and catechins in green tea are equally effective antioxidants. *J. Nutr.* 131(9):2248–51.
57. Loke, W. M., Proudfoot, J. M., Hodgson, J. M., et al. 2010. Specific dietary polyphenols attenuate atherosclerosis in apolipoprotein E–knockout mice by alleviating inflammation and endothelial dysfunction. *Arterioscler Thromb. Vasc. Biol.* 30(4):749–57.
58. Sugatani, J., Fukazawa, N., Ujihara, K., et al. 2004. Tea polyphenols inhibit acetyl-CoA: 1-alkyl-sn-glycero-3-phosphocholine acetyltransferase (a key enzyme in platelet-activating factor biosynthesis) and platelet-activating factor-induced platelet aggregation. *Int. Arch. Allergy Immunol.* 134(1):17–28.
59. Łuczaj, W., Zapora, E., Szczepański, M., Wnuczko, K. and Skrzydlewska, E. 2009. Polyphenols action against oxidative stress formation in endothelial cells. *Acta Pol. Pharm.* 66(6):617–24.

60. Lorenz, M., Urban, J., Engelhardt, U., Baumann. G., Stangl, K. and Stangl, V. 2009. Green and black tea are equally potent stimuli of NO production and vasodilation: New insights into tea ingredients involved. *Basic Res. Cardiol.* 104(1):100–10.
61. Jankun, J. and Skotnicka, M., Łysiak-Szydłowska, W., Al-Senaidy, A. and Skrzypczak-Jankun, E. 2011. Diverse inhibition of plasminogen activator inhibitor type 1 by theaflavins of black tea. *Int. J. Mol. Med.* 27(4):525–9.
62. Grassi, D., Aggio, A., Onori, L., et al. 2008. Tea, flavonoids, and nitric oxide-mediated vascular reactivity. *J. Nutr.* 138(8):1554S–1560S.
63. Geleijnse, J. M., Launer, L. J., van der Kuip, D. A., Hofman, A and Witteman, J. C. 2002. Inverse association of tea and flavonoid intakes with incident myocardial infarction: The Rotterdam Study. *Am. J. Clin. Nutr.* 75(5):880–6.
64. Widlansky, M. E., Duffy, S. J., Hamburg, N. M., et al. 2005. Effects of black tea consumption on plasma catechins and markers of oxidative stress and inflammation in patients with coronary artery disease. *Free Radical. Biol. Med.* 38(4):499–506.
65. Nishiumi, S., Bessyo, H., Kubo, M., et al. 2010. Green and black tea suppress hyperglycemia and insulin resistance by retaining the expression of glucose transporter 4 in muscle of high-fat diet-fed C57BL/6J mice. *J. Agric. Food Chem.* 58(24):12916–23.
66. Uchiyama, S., Taniguchi, Y., Saka, A., Yoshida, A. and Yajima, H. 2011. Prevention of diet-induced obesity by dietary black tea polyphenols extract *in vitro* and *in vivo*. *Nutr.* 27(3):287–92.
67. Yang, M.-H., Wang, C.-H. and Chen, H.-L. 2001. Green, oolong and black tea extracts modulate lipid metabolism in hyperlipidemia rats fed high-sucrose diet. *J. Nutr. Biochem.* 12(1):14–20.
68. Majumdar, S., Karmakar, S., Maiti, A., et al. 2011. Arsenic-induced hepatic mitochondrial toxicity in rats and its amelioration by dietary phosphate. *Environ. Toxicol. Pharmacol.* 31(1):107–18.
69. Luo, X.-Y., Takahara, T., Hou, J., et al. 2012. Theaflavin attenuates ischemia–reperfusion injury in a mouse fatty liver model. *Biochem. Biophys. Res. Commun.* 417(1):287–93.
70. Lin, C.-L., Huang, H.-C. and Lin, J.-K. 2007. Theaflavins attenuate hepatic lipid accumulation through activating AMPK in human HepG2 cells. *J. Lipid Res.* 48(11):2334–43.
71. Vermeer, M. A., Mulder, T. P. and Molhuizen, H. O. 2008. Theaflavins from black tea, especially theaflavin-3-gallate, reduce the incorporation of cholesterol into mixed micelles. *J. Agric. Food Chem.* 56(24):12031–6.
72. Ikeda, I., Yamahira, T., Kato, M. and Ishikawa, A. 2010. Black-tea polyphenols decrease micellar solubility of cholesterol *in vitro* and intestinal absorption of cholesterol in rats. *J. Agric. Food Chem.* 58(15):8591–5.
73. Kobayashi, M., Ichitani, M., Suzuki, Y., et al. 2009. Black-tea polyphenols suppress postprandial hypertriacylglycerolemia by suppressing lymphatic transport of dietary fat in rats. *J. Agric. Food Chem.*57(15):7131–6.
74. Gauci, A. J., Caruana, M., Giese, A., Scerri, C. and Vassallo, N. 2011. Identification of polyphenolic compounds and black tea extract as potent inhibitors of lipid membrane destabilization by Aβ42 aggregates. *J. Alzheimers Dis.* 27(4):767–79.
75. Skrzypczak-Jankun, E and Jankun J. 2010. Theaflavin digallate inactivates plasminogen activator inhibitor: Could tea help in Alzheimer's disease and obesity? *Int. J. Mol. Med.* 26(1):45–50.
76. Grelle, G., Otto, A., Lorenz, M., Frank, R. F., Wanker, E. E. and Bieschke, J. 2011. Black tea theaflavins inhibit formation of toxic amyloid-β and α-synuclein fibrils. *Biochem.* 50(49):10624–36.
77. Tan, L. C., Koh, W.-P., Yuan, J.-M., et al. 2007. Differential effects of black versus green tea on risk of Parkinson's disease in the Singapore Chinese Health Study. *Am. J. Epidemiol.* 167(5):553–60.
78. Chaturvedi, R., Shukla, S., Seth, K., et al. 2006. Neuroprotective and neurorescue effect of black tea extract in 6-hydroxydopamine-lesioned rat model of Parkinson's disease. *Neurobiol. Dis.* 22(2):421–34.
79. Anandhan, A., Tamilselvam, K., Radhiga, T., Rao, S., Essa, M. M. and Manivasagam, T. 2012. Theaflavin, a black tea polyphenol, protects nigral dopaminergic neurons against chronic MPTP/probenecid induced Parkinson's disease. *Brain Res.* 1433:104–13.
80. Peng, C., Chan, H. Y. E., Li, Y. M., Huang, Y. and Chen, Z. Y. 2009. Black tea theaflavins extend the lifespan of fruit flies. *Exp Gerontol.* 44(12):773–83.

81. Cameron, A. R., Anton, S., Melville, L., et al. 2008. Black tea polyphenols mimic insulin/insulin-like growth factor-1 signalling to the longevity factor FOXO1a. *Aging cell.* 7(1):69–77.
82. Zhang, L., Zhang, Z.-Z., Lu, Y.-N., Zhang, J.-S. and Preedy, V. R., 2013. L-Theanine from green tea: Transport and effects on health. In: Preedy, V. R. (Ed.), *Tea in Health and Disease Prevention*, Academic Press, Waltham, pp. 425–35.
83. Uchiyama, S., Taniguchi, Y., Saka, A., Yoshida, A. and Yajima, H. 2011. Prevention of diet-induced obesity by dietary black tea polyphenols extract *in vitro* and *in vivo*. *Nutr.* 27(3):287–92.
84. Li, S., Lo, C.-Y., Pan, M.-H., Lai, C.-S. and Ho, C.-T. 2013. Black tea: Chemical analysis and stability. *Food Funct.* 4(1):10–18.
85. Zhang, X., Zhu, X., Sun, Y., et al. 2013. Fermentation *in vitro* of EGCG, GCG and EGCG3′Me isolated from Oolong tea by human intestinal microbiota. *Food Res. Int.* 54(2):1589–95.
86. Kaneko, S., Kenji Kumazawa, H. M. and Hofmann, T. 2006. Molecular and sensory studies on the umami taste of Japanese green tea. *J. Agric. Food Chem.* 54(7):2688–94.
87. Yokogoshi, H., Kato, Y., Sagesaka, Y. M., Takihara-Matsuura, T., Kakuda, T. and Takeuchi, N. 1995. Reduction effect of theanine on blood pressure and brain 5-hydroxyindoles in spontaneously hypertensive rats. *Biosci. Biotechnol. Biochem.* 59(4):615–18.
88. Kakuda, T. 2002. Neuroprotective effects of the green tea components theanine and catechins. *Biol. Pharm. Bull.* 25(12):1513.
89. Nathan, P. J., Lu, K., Gray, M. and Oliver, C. 2006. The neuropharmacology of L-theanine(N-ethyl-L-glutamine): A possible neuroprotective and cognitive enhancing agent. *J. Herb. Pharmacother.* 6(2):21–30.
90. Sagesaka, Y. M., Uemura, T., Watanabe, N., Sakata, K. and Uzawa, J. 1994. A new glucuronide saponin from tea leaves (*Camellia sinensis var. sinensis*). *Biosci. Biotechnol. Biochem.* 58(11):2036–40.
91. Murakami, T., Nakamura, J., Kageura, T., Matsuda, H. and Yoshikawa, M. 2000. Bioactive saponins and glycosides. XVII. Inhibitory effect on gastric emptying and accelerating effect on gastrointestinal transit of tea saponins: Structures of assamsaponins F, G, H, I, and J from the seeds and leaves of the tea plant. *Chem. Pharm. Bull. (Tokyo)* 48(11):1720–5.
92. Kobayashi, K., Teruya, T., Suenaga, K., Matsui, Y., Masuda, H. and Kigoshi, H. 2006. Isotheasaponins B1–B3 from *Camellia sinensis var. sinensis* tea leaves. *Phytochem.* 67(13):1385–9.
93. Morikawa, T., Nakamura, S., Kato, Y., Muraoka, O., Matsuda, H. and Yoshikawa, M. 2007. Bioactive saponins and glycosides. XXVIII. New triterpene saponins, foliatheasaponins I, II, III, IV, and V, from Tencha (the leaves of *Camellia sinensis*). *Chem. Pharm. Bull. (Tokyo)* 55(2):293–8.
94. Wang, K., Liu, F., Liu, Z., et al. 2011. Comparison of catechins and volatile compounds among different types of tea using high performance liquid chromatograph and gas chromatograph mass spectrometer. *Int. J. Food Sci. Tech.* 46(7):1406–12.
95. Shimoda, M., Shigematsu, H., Shiratsuchi, H. and Osajima, Y. 1995. Comparison of the odor *concentrates* by SDE and adsorptive column method from green tea infusion. *J. Agric. Food Chem.* 43(6):1616–20.
96. And, C. S. and Schieberle, P. 2006. Characterization of the key aroma compounds in the beverage prepared from Darjeeling black tea: Quantitative differences between tea leaves and infusion. *J. Agric. Food Chem.* 54(3):916–24.
97. Lian, M., Shi-Dong, L., Yuan-Shuang, W. U., Zhou, J. S., Wang, C. and Meng, Q. X. 2015. Analysis of volatile compounds of four kinds of black tea from China. *J. Trop. Subtrop. Bot.* 03:301–9.
98. Mahanta, P. K., Hazarika, M. and Takeo, T. 1985. Flavour volatiles and lipids in various components of tea shoots *Camellia sinensis* (L.) O. Kuntze. *J. Sci. Food Agric.* 36(11):1130–2.
99. Bhuyan, L. P., Tamuly, P. and Mahanta, P. K. 1991. Lipid content and fatty acid composition of tea shoot and manufactured tea. *J. Agric. Food Chem.* 39(6):1159–62.
100. Bhuyan, L. P. and Mahanta, P. K. 2010. Studies on fatty acid composition in tea *Camellia sinensis*. *J. Sci. Food Agric.* 46(3):325–30.
101. Mahanta, P. K. and Hazarika, M. 2010. Chlorophylls and degradation products in orthodox and CTC black teas and their influence on shade of colour and sensory quality in relation to thearubigins. *J. Sci. Food. Agric.* 36(11):1133–9.

102. And, Y. S. and Shioi, Y. 2003. Identification of chlorophylls and carotenoids in major teas by high-performance liquid chromatography with photodiode array detection. *J. Agric. Food Chem.* 51(18):5307–14.
103. Kuroki, M., Shioi, Y. and Sasa, T. 1981. Purification and properties of soluble chlorophyllase from tea leaf sprouts. *Plant Cell Physiol.* 22(4):717–25.
104. Obanda, M. and Owuor, P. O. 1995. Impact of shoot maturity on chlorophyll content, composition of volatile flavour compounds and plain black tea chemical quality parameters of clonal leaf. *J. Sci. Food Agric.* 69(4):529–34.
105. Van Lelyveld, L., Smith, B., Fraser, C. and Visser, G. 1990. Variation in quality of certain tea clones with respect to chlorophyll, theaflavin content and total colour value. *South Afr. J. Plant Soil* 7(4):226–9.
106. Van Lelyveld, L. and Smith, B. 1989. The association between residual chlorophyll and grassy taste in black tea. *South Afr. J. Plant Soil* 6(4):280–1.
107. Suzuki, Y. and Shioi, Y. 2003. Identification of chlorophylls and carotenoids in major teas by high-performance liquid chromatography with photodiode array detection. *J. Agric. Food Chem.* 51(18):5307–14.
108. Sanderson, G. W. and Graham, H. N. 1973. Formation of black tea aroma. *J. Agric. Food Chem.* 21(4):576–85.
109. Sanderson, G. W., Co, H. and Gonzalez, J. G. 1971. Biochemistry of tea fermentation: The role of carotenes in black tea aroma formation. *J. Food Sci.* 36(2):231–6.
110. Renold, W., Näf-Müller, R., Keller, U., Willhalm, B. and Ohloff, G. 1974. An investigation of the tea aroma. Part I. New volatile black tea constituents. *Helv. Chim. Acta.* 57(5):1301–8.
111. Dalluge, J. J. and Nelson, B. C. 2000. Determination of tea catechins. *J. Chromatogr. A* 881(1–2):411–24.
112. Li, P., Song, X., Shi, X., Li, J. and Ye, C. 2008. An improved HPLC method for simultaneous determination of phenolic compounds, purine alkaloids and theanine in Camellia species. *J. Food Comp. Anal.* 21(7):559–63.
113. Zuo, Y., Chen, H. and Deng, Y. 2002. Simultaneous determination of catechins, caffeine and gallic acids in green, oolong, black and pu-erh teas using HPLC with a photodiode array detector. *Talanta* 57(2):307–16.
114. Wang, D., Lu, J., Miao, A., Xie, Z. and Yang, D. 2008. HPLC-DAD-ESI-MS/MS analysis of polyphenols and purine alkaloids in leaves of 22 tea cultivars in China. *J. Food Comp. Anal.* 21(5):361–9.
115. Pongsuwan, W., Bamba, T., Harada, K., Yonetani, T., Kobayashi, A. and Fukusaki, E. 2008. High-throughput technique for comprehensive analysis of Japanese green tea quality assessment using ultra-performance liquid chromatography with time-of-flight mass spectrometry (UPLC/TOF MS). *J. Agric. Food. Chem.* 56(22):10705–8.
116. Zhao, Y., Chen, P., Lin, L., Harnly, J., Yu, L. L. and Li, Z. 2011. Tentative identification, quantitation, and principal component analysis of green pu-erh, green, and white teas using UPLC/DAD/MS. *Food Chem.* 126(3):1269–77.
117. Jiang, X., Xia, Z., Wei, W. and Gou, Q. 2009. Direct UV detection of underivatized amino acids using capillary electrophoresis with online sweeping enrichment. *J. Sep. Sci.* 32(11):1927–33.
118. Hsieh, M. M. and Chen, S. M. 2007. Determination of amino acids in tea leaves and beverages using capillary electrophoresis with light-emitting diode-induced fluorescence detection. *Talanta* 73(2):326–31.
119. Wright, L. P., Mphangwe, N. I. K., Nyirenda, H. E. and Apostolides, Z. 2002. Analysis of the theaflavin composition in black tea (*Camellia sinensis*) for predicting the quality of tea produced in Central and Southern Africa. *J. Sci. Food Agric.* 82(5):517–25.
120. Horie, H., Mukai, T. and Kohata, K. 1997. Simultaneous determination of qualitatively important components in green tea infusions using capillary electrophoresis. *J. Chromatogr. A* 758(2):332–5.

121. Kodama, S., Ito, Y., Nagase, H., et al. 2007. Usefulness of catechins and caffeine profiles to determine growing areas of green tea leaves of a single variety, Yabukita, in Japan. *J. Health Sci.* 53(4):491–5.
122. Beelan, L. and Choonnam, O. 2000. Comparative analysis of tea catechins and theaflavins by high-performance liquid chromatography and capillary electrophoresis. *J. Chromatogr. A* 881(1–2):439–47.
123. Sereshti, H., Samadi, S. and Jalali-Heravi, M. 2013. Determination of volatile components of green, black, oolong and white tea by optimized ultrasound-assisted extraction-dispersive liquid–liquid microextraction coupled with gas chromatography. *J. Chromatogr. A* 1280(4):1–8.
124. Lv, H. P., Zhong, Q. S., Lin, Z., Wang, L., Tan, J. F. and Guo, L. 2012. Aroma characterisation of Pu-erh tea using headspace-solid phase microextraction combined with GC/MS and GC–olfactometry. *Food Chem.* 130(4):1074–81.
125. Pang, G. F., Fan, C. L., Zhang, F., et al. 2011. High-throughput GC/MS and HPLC/MS/MS techniques for the multiclass, multiresidue determination of 653 pesticides and chemical pollutants in tea. *J. AOAC Int.* 94(4):1253–96.
126. Schurek, J., Portolés, T., Hajslova, J., Riddellova, K. and Hernández, F. 2008. Application of head-space solid-phase microextraction coupled to comprehensive two-dimensional gas chromatography-time-of-flight mass spectrometry for the determination of multiple pesticide residues in tea samples. *Anal. Chim. Acta.* 611(2):163–72.
127. Ochiai, N., Sasamoto, K., Kanda, H., et al. 2005. Optimization of a multi-residue screening method for the determination of 85 pesticides in selected food matrices by stir bar sorptive extraction and thermal desorption GC-MS. *J. Sep. Sci.* 28(9–10):1083.
128. Pokinda, O., Martin, O., Hastingse, N. and Wilson, M. 2008. Influence of region of production on clonal black tea chemical characteristics. *Food Chem.* 108(1):263–71.
129. Pongsuwan, W., Fukusaki, E., Bamba, T., Yonetani, T., Yamahara, T. and Kobayashi, A. 2007. Prediction of Japanese green tea ranking by gas chromatography/mass spectrometry-based hydrophilic metabolite fingerprinting. *J. Agric. Food Chem.* 55(2):231–6.
130. Reich, E., Schibli, A., Widmer, V., Jorns, R., Wolfram, E. and DeBatt, A. 2006. HPTLC methods for identification of green tea and green tea extract. *J. Liq. Chromatogr. Rel. Technol.* 29(14):2141–51.
131. Le, G. G., Colquhoun, I. J. and Defernez, M. 2004. Metabolite profiling using (1)H NMR spectroscopy for quality assessment of green tea, *Camellia sinensis* (L.). *J. Agric. Food Chem.* 52(4):692–700.
132. Tarachiwin, L., Ute, K., Kobayashi, A. and Fukusaki, E. 2007. 1H NMR based metabolic profiling in the evaluation of Japanese green tea quality. *J. Agric. Food Chem.* 55(23):9330.
133. Fujiwara, M., Ando, I. and Arifuku, K. 2006. Multivariate analysis for 1H-NMR spectra of two hundred kinds of tea in the world. *Anal. Sci.* 22(10):1307–14.
134. Son, H. S., Kim, K. M., Van den Berg, F., et al. 2008. 1H nuclear magnetic resonance-based metabolomic characterization n of wines by grape varieties and production areas. *J. Agric. Food Chem.* 56(17):8007–16.
135. Son, H. S., Hwang, G. S., Kim, K. M., et al. 2009. Metabolomic studies on geographical grapes and their wines using 1H NMR analysis coupled with multivariate statistics. *J. Agric. Food Chem.* 57(4):1481–90.
136. Lee, J. E., Hwang, G. S., Van, D. B. F., Lee, C. H. and Hong, Y. S. 2009. Evidence of vintage effects on grape wines using 1H NMR-based metabolomic study. *Anal. Chim. Acta.* 648(1):71–6.
137. Jian, Z., Hao, C., Wei, H., Wang, L. Y., Xu, L. and Lu, W. Y. 2009. Identification of geographical indication tea with Fisher's discriminant classification and principal components analysis. *J. Near Infrared Spectrosc.* 17(3):159–64.
138. Zhao, J., Chen, Q., Huang, X. and Fang, C. H. 2006. Qualitative identification of tea categories by near infrared spectroscopy and support vector machine. *J. Pharm. Biomed. Anal.* 41(4):1198–204.
139. Chen, Q., Zhao, J., Fang, C. H. and Wang, D. 2007. Feasibility study on identification of green, black and Oolong teas using near-infrared reflectance spectroscopy based on support vector machine (SVM). *Spectrochim. Acta A Mol. Biomol. Spectrosc.* 66(3):568–74.

140. Chen, Q., Zhao, J., Zhang, H. and Wang, X. 2006. Feasibility study on qualitative and quantitative analysis in tea by near infrared spectroscopy with multivariate calibration. *Anal. Chim. Acta.* 572(1):77–84.
141. Luypaert, J., Zhang, M. H. and Massart, D. L. 2003. Feasibility study for the use of near infrared spectroscopy in the qualitative and quantitative analysis of green tea, *Camellia sinensis* (L.). *Anal. Chim. Acta.* 478(2):303–12.
142. Schulz, H., Engelhardt, U. H., Wegent, A., Drews, H. and Lapczynski, S. 1999. Application of near-infrared reflectance spectroscopy to the simultaneous prediction of alkaloids and phenolic substances in green tea leaves. *J. Agric. Food Chem.* 47(12):5064–7.
143. Li, X. and He, Y. 2008. Discriminating varieties of tea plant based on Vis/NIR spectral characteristics and using artificial neural networks. *Biosys. Eng.* 99(3):313–21.
144. Liu, F., Ye, X. J., He, Y. and Wang, L. 2009. Application of visible/near infrared spectroscopy and chemometric calibrations for variety discrimination of instant milk teas. *J. Food Eng.* 93(2):127–33.
145. Chen, Q., Zhao, J., Chaitep, S. and Guo, Z. 2009. Simultaneous analysis of main catechins contents in green tea (*Camellia sinensis* (L.)) by Fourier transform near infrared reflectance (FT-NIR) spectroscopy. *Food Chem.* 113(4):1272–7.
146. Palacios-Morillo, A., Alcazar, A., de Pablos, F. and Jurado, J. M. 2013. Differentiation of tea varieties using UV-Vis spectra and pattern recognition techniques. *Spectrochim. Acta A Mol. Biomol. Spectrosc.* 103:79–83.
147. Martelo-Vidal, M. J. and Vazquez, M. 2014. Determination of polyphenolic compounds of red wines by UV-VIS-NIR spectroscopy and chemometrics tools. *Food Chem.* 158:28–34.
148. Li, K., Shi, X., Yang, X., Wang, Y., Ye, C. and Yang, Z. 2012. Antioxidative activities and the chemical constituents of two Chinese teas, *Camellia* kucha and *C. ptilophylla*. *Int. J. Food Sci. Tech.* 47(5):1063–71.
149. Islam, G. M. R., Uddin, M. G., Rahman, M. M. and Yousuf, A. 2013. Caffeine and total polyphenol contents of market tea cultivated and processed in Bangladesh. *Malaysian J. Nutr.* 19(1):143–7.
150. Atomssa, T. and Gholap, A. V. 2011. Characterization of caffeine and determination of caffeine in tea leaves using UV-visible spectrometer. *Urología.* 38(3):1–12.
151. López-Martínez, L., López-de-Alba, P. L., García-Campos, R. and De León-Rodríguez, L. M. 2003. Simultaneous determination of methylxanthines in coffees and teas by UV-Vis spectrophotometry and partial least squares. *Anal. Chim. Acta.* 493(1):83–94.
152. Akuli, A., Pal, A., Bej, G., et al. 2016. A machine vision system for estimation of theaflavins and thearubigins in orthodox black tea. *Int. J on Smart Sensing and Intelligent Systems.*;9(2):709–31.
153. Seenivasan, S., Manikandan, N., Muraleedharan, N. N. and Selvasundaram, R. 2008. Heavy metal content of black teas from south India. *Food Control.* 19(8):746–9.
154. Fernandez, P. L., Pablos, F., Martin, M. J. and Gonzalez, A. G. 2002. Multi-element analysis of tea beverages by inductively coupled plasma atomic emission spectrometry. *Food Chem.* 76(4):483–9.
155. Matsuura, H., Hokura, A., Katsuki, F., Itoh, A. and Haraguchi, H. 2001. Multielement determination and speciation of major-to-trace elements in black tea leaves by ICP-AES and ICP-MS with the aid of size exclusion chromatography. *Anal. Sci.* 17(3):391–8.
156. Feng, H., Wang, T. and Li, S. F. Y. 2003. Sensitive determination of trace-metal elements in tea with capillary electrophoresis by using chelating agent 4-(2-pyridylazo) resorcinol (PAR). *Food Chem.* 81(4):607–11.
157. Xie, M., Bohlen, A. V., Klockenkämper, R., Jian, X. and Günther, K. 1998. Multielement analysis of Chinese tea (*Camellia sinensis*) by total-reflection X-ray fluorescence. *Eur. Food Res. Technol.* 207(1):31–8.
158. Salahinejad, M. and Aflaki, F. 2010. Toxic and essential mineral elements content of black tea leaves and their tea infusions consumed in Iran. *Biol. Trace Elem. Res.* 134(1):109–17.
159. Boogers, I., Plugge, W., Stokkermans, Y. Q. and Duchateau, A. L. L. 2008. Ultra-performance liquid chromatographic analysis of amino acids in protein hydrolysates using an automated pre-column derivatisation method. *J. Chromatogr. A* 1189(1–2):406–9.

160. Kaspar, H., Dettmer, K., Gronwald, W. and Oefner, P. J. 2008. Automated GC–MS analysis of free amino acids in biological fluids. *J. Chromatogr. B - Analyt. Technol. Biomed. Life Sci.* 870(2):222–32.
161. Lin, C., Qi, C., Zhang, Z. Z. and Wan, X. C. 2009. A novel colorimetric determination of free amino acids content in tea infusions with 2,4-dinitrofluorobenzene. *J Food Comp. Anal.* 22(2):137–41.
162. Le, A., Ng, A., Kwan, T., Cusmano-Ozog, K. and Cowan, T. M. 2014. A rapid, sensitive method for quantitative analysis of underivatized amino acids by liquid chromatography-tandem mass spectrometry (LC-MS/MS). *J. Chromatogr. B.* 944:166–74.
163. Pinto, M. C., de Paiva, M. J., Oliveiralima, O. C., et al. 2014. Neurochemical study of amino acids in rodent brain structures using an improved gas chromatography-mass spectrometry method. *J. Chem. Neuroanat.* 55(2):24–37.
164. Ding, Y., Yu, H. and Mou, S. 2002. Direct determination of free amino acids and sugars in green tea by anion-exchange chromatography with integrated pulsed amperometric detection. *J. Chromatogr. A.* 982(2):237–44.
165. Thippeswamy, R., Mallikarjun Gouda, K. G., Rao, D. H., Martin, Asha and Gowda, L. R. 2006. Determination of theanine in commercial tea by liquid chromatography with fluorescence and diode array ultraviolet detection. *J. Agric. Food Chem.* 54(19):7014–19.
166. Feng, H. T. and Li, S. F. Y. 2002. Determination of five toxic alkaloids in two common herbal medicines with capillary electrophoresis. *J. Chromatogr. A* 973(1–2):243.
167. Zhu, X., Chen, B., Ma, M., et al. 2004. Simultaneous analysis of theanine, chlorogenic acid, purine alkaloids and catechins in tea samples with the help of multi-dimension information of on-line high performance liquid chromatography/electrospray-mass spectrometry. *J. Pharm. Biomed. Anal.* 34(3):695–704.
168. Sereshti, H. and Samadi, S. 2014. A rapid and simple determination of caffeine in teas, coffees and eight beverages. *Food Chem.* 158(5):8–13.
169. Larger, P. J., Jones, A. D. and Dacombe, C. 1998. Separation of tea polyphenols using micellar electrokinetic chromatography with diode array detection. *J. Chromatogr. A* 799(1–2):309–20.
170. Khokhar, S. and Magnusdottir, S. G. 2002. Total phenol, catechin, and caffeine contents of teas commonly consumed in the United kingdom. *J. Agric. Food Chem.* 50(3):565–70.
171. Filip, R. and Ferraro, G. E. 2003. Researching on new species of 'Mate': *Ilex brevicuspis*: Phytochemical and pharmacology study. *Eur. J. Nutr.* 42(1):50–4.
172. Liang, Y., Lu, J., Zhang, L., Wu, S. and Wu, Y. 2003. Estimation of black tea quality by analysis of chemical composition and colour difference of tea infusions. *Food Chem.* 80(2):283–90.
173. Ehala, S., Merike Vaher, A. and Kaljurand, M. 2005. Characterization of phenolic profiles of Northern European berries by capillary electrophoresis and determination of their antioxidant activity. *J. Agric. Food Chem.* 53(16):6484–90.
174. Nováková, L., Vildová, A., Mateus, J. P., Gonçalves, T. and Solich, P. 2010. Development and application of UHPLC-MS/MS method for the determination of phenolic compounds in Chamomile flowers and Chamomile tea extracts. *Talanta.* 82(4):1271–80.
175. Nie, S., Xie, M., Fu, Z., Wan, Y. and Yan, A. 2008. Study on the purification and chemical compositions of tea glycoprotein. *Carbohydr. Polym.* 71(4):626–33.
176. Pytlakowska, K., Kita, A., Janoska, P., Połowniak, M. and Kozik, V. 2012. Multi-element analysis of mineral and trace elements in medicinal herbs and their infusions. *Food Chem.* 135(2):494–501.
177. Froes, R. E. S., Neto, W. B., Beinner, M. A., Nascentes, C. C. and Silva, J. B. B. D. 2014. Determination of inorganic elements in teas using inductively coupled plasma optical emission spectrometry and classification with exploratory analysis. *Food Anal. Methods* 7(3):540–6.
178. Szymczycha-Madeja, A., Welna, M. and Pohl, P. 2014. Simple and fast sample preparation procedure prior to multi-element analysis of slim teas by ICP OES. *Food Anal. Methods* 7(10):2051–63.
179. Pazrodríguez, B., Domínguezgonzález, M. R., Aboalsomoza, M. and Bermejobarrera, P. 2015. Application of high resolution-continuum source flame atomic absorption spectrometry (HR-CS FAAS): Determination of trace elements in tea and tisanes. *Food Chem.* 170:492–500.

180. Szymczycha-Madeja, A., Welna, M. and Pohl, P. 2015. Determination of essential and non-essential elements in green and black teas by FAAS and ICP OES simplified–multivariate classification of different tea products. *Microchem. J.* 121:122–9.
181. Lapornik, B., Prosek, M. and Wondra, A. G. 2005. Comparison of extracts prepared from plant by-products using different solvents and extraction time. *J. Food Eng.* 71(2):214–22.
182. Harbowy, M. E., Balentine, D. A., Davies, D. A. P. and Cai, D. Y. 1997. Tea chemistry. *Crit. Rev. Plant Sci.* 16(5):415–80.
183. Temple, C. M. and Clifford, M. N. 1997. The stability of theaflavins during HPLC analysis of a decaffeinated aqueous tea extract. *J. Sci. Food Agric.* 74(4):536–40.
184. Waterhouse, A. L. 2003. *Determination of Total Phenolics*. In: Wrolstad, R. E., Acree, T. E., Decker, E. A., Penner, M. H., Reid, D. S., Schwartz, S. J., hoemaker, C. F. Smith, S. D. M. and Sporns, P. (Eds), *Food Analytical Chemistry*, John Wiley & Sons, Inc., New York, pp. I1.1.1– I1.1.8.
185. Nishitani, E. and Sagesaka, Y. M. 2004. Simultaneous determination of catechins, caffeine and other phenolic compounds in tea using new HPLC method. *J. Food Comp. & Anal.* 17(5):675–85.
186. Dou, J., Lee, V. S. Y., Tzen, J. T. C. and Lee, M. R. 2007. Identification and comparison of phenolic compounds in the preparation of oolong tea manufactured by semifermentation and drying processes. *J. Agric. Food Chem.* 55(18):7462–8.
187. Canas, S., Belchior, A. P., Spranger, M. I. and Brunodesousa, R. 2010. HPLC method for the quantification of phenolic acids, phenolic aldehydes, coumarins and furanic derivatives in different kinds of toasted wood used for the ageing of brandies. *Anal. Methods.* 3(1):186–91.
188. Wang, Y., Yang, X., Li, K., et al. 2010. Simultaneous determination of theanine, gallic acid, purine alkaloids, catechins, and theaflavins in black tea using HPLC. *Int. J. Food Sci. Technol.* 45(6):1263–9.
189. El-Shahawi, M. S., Hamza, A., Bahaffi, S. O., Al-Sibaai, A. A. and Abduljabbar, T. N. 2012. Analysis of some selected catechins and caffeine in green tea by high performance liquid chromatography. *Food Chem.* 134(4):2268–75.
190. Riverocruz, B. 2014. Simultaneous quantification by HPLC of the phenolic compounds for the crude drug of *Prunus serotina subsp. capuli. Pharm. Biol.* 52(8):1015–20.
191. Bae, I. K., Ham, H. M., Jeong, M. H., Kim, D. H. and Kim, H. J. 2015. Simultaneous determination of 15 phenolic compounds and caffeine in teas and mate using RP-HPLC/UV detection: Method development and optimization of extraction process. *Food Chem.* 172:469–475.
192. Ignat, I., Volf, I. and Popa, V. I. 2011. A critical review of methods for characterisation of polyphenolic compounds in fruits and vegetables. *Food Chem.* 126(4):1821–35.
193. Oh, Y. S., Lee, J. H., Yoon, S. H., et al. 2008. Characterization and quantification of anthocyanins in grape juices obtained from the grapes cultivated in Korea by HPLC/DAD, HPLC/MS, and HPLC/MS/MS. *J. Food Sci.* 73(5):C378–89.
194. Del Rio, D., Stewart, A. J., Mullen, W., Burns, J., Lean, M. E. J., Brighenti, F. and Crozier, A. 2004. HPLC-MSn analysis of phenolic compounds and purine alkaloids in green and black tea. *J. Agric. Food Chem.* 52(10):2807–15.
195. Arce, L., Ríos, A. and Valcárcel, M. 1998. Determination of anti-carcinogenic polyphenols present in green tea using capillary electrophoresis coupled to a flow injection system. *J. Chromatogr. A* 827(1):113–20.
196. Rana, A. and Singh, H. P. 2012. A rapid HPLC-DAD method for analysis of theaflavins using c12 as stationary phase. *J. Liq. Chromatogr. Rel. Technol.* 35(16):2272–9.
197. Wedzicha, B., Lo, M. and Donovan, T. 1990. Counter-current chromatography of black tea infusions. *J. Chromatogr. A* 505(2):357–64.
198. Degenhardt, A., Engelhardt, U. H., Wendt, A.-S. and Winterhalter, P. 2000. Isolation of black tea pigments using high-speed countercurrent chromatography and studies on properties of black tea polymers. *J. Agric. Food Chem.* 48(11):5200–5.
199. Cao, X., Lewis, J. R. and Ito, Y. 2004. Application of high-speed countercurrent chromatography to the separation of black tea theaflavins. *J. Liq. Chromatogr. Rel. Technol.* 27(12):1893–902.

200. Powell, C., Clifford, M. N., Opie, S. C., Ford, M. A., Robertson, A. and Gibson C. L. 1993. Tea cream formation: The contribution of black tea phenolic pigments determined by H.P.L.C.. *J. Sci. Food Agric.* 63(1):77–86.
201. Menet, M.-C., Sang, S., Yang, C. S., Ho, C.-T. and Rosen, R. T. Analysis of theaflavins and thearubigins from black tea extract by MALDI-TOF mass spectrometry. *J. Agric. Food Chem.* 52(9):2455–61.
202. Kuhnert, N., Drynan, J. W., Obuchowicz, J., Clifford, M. N. and Witt, M. 2010. Mass spectrometric characterization of black tea thearubigins leading to an oxidative cascade hypothesis for thearubigin formation. *Rapid Commun. Mass Spectrom.* 24(23):3387–404.
203. Drynan, J. W., Clifford, M. N., Obuchowicz, J. and Kuhnert, N. 2012. MALDI-TOF mass spectrometry: Avoidance of artifacts and analysis of caffeine-precipitated SII thearubigins from 15 commercial black teas. *J. Agric. Food Chem.* 60(18):4514–25.
204. Chen, H. X., Zhang, M. and Xie, B. 2005. Components and antioxidant activity of polysaccharide conjugate from green tea. *Food Chem.* 90(1–2):17–21.
205. Reto, M., Figueira, M. E., Filipe, H. M. and Almeida, C. M. M. 2007. Analysis of vitamin K in green tea leafs and infusions by SPME-GC-FID. *Food Chem.* 100(1):405–11.
206. Shrivas, K. and Wu, H. F. 2007. Rapid determination of caffeine in one drop of beverages and foods using drop-to-drop solvent microextraction with gas chromatography/mass spectrometry. *J. Chromatogr. A* 1170(1–2):9–14.
207. Ochiai, N., Sasamoto, K., Hoffmann, A. and Okanoya, K. 2012. Full evaporation dynamic headspace and gas chromatography–mass spectrometry for uniform enrichment of odor compounds in aqueous samples. *J. Chromatogr. A.* 1240(11):59–68.
208. Horie, H., Nesumi, A., Ujihara, T. and Kohata, K. 2002. Rapid determination of caffeine in tea leaves. *J. Chromatogr. A* 942(1–2):271–3.
209. Al-Othman, Z. A., Aqel, A., Alharbi, M. K. E., Badjah-Hadj-Ahmed, A. Y. and Al-Warthan, A. A. 2012. Fast chromatographic determination of caffeine in food using a capillary hexyl methacrylate monolithic column. *Food Chem.* 132(4):2217–23.
210. Kerio, L. C., Wachira, F. N., Wanyoko, J. K. and Rotich, M. K. 2012. Characterization of anthocyanins in Kenyan teas: Extraction and identification. *Food Chem.* 131(1):31–8.
211. Wang, L., Gong, L. H., Chen, C. J., Han, H. B. and Li, H. H. 2016. Column–chromatographic extraction and separation of polyphenols, caffeine and theanine from green tea. *J. Crohns Colitis.* 10(8):873–85.
212. Chen, G., Wang, Y., Song, W., Zhao, B. and Dou, Y. 2012. Rapid and selective quantification of l -theanine in ready-to-drink teas from Chinese market using SPE and UPLC-UV. *Food Chem.* 135(2):402–7.
213. Krishnan, R. and Maru, G. B. 2006. Isolation and analyses of polymeric polyphenol fractions from black tea. *Food Chem.* 94(3):331–40.
214. Tanizawa, Y., Abe, T. and Yamada, K. 2007. Black tea stain formed on the surface of teacups and pots. Part 1 – Study on the chemical composition and structure. *Food Chem.* 103(1):1–7.
215. Jafari, M. T., Rezaei, B. and Javaheri, M. 2011. A new method based on electrospray ionisation ion mobility spectrometry (ESI-IMS) for simultaneous determination of caffeine and theophylline. *Food Chem.* 126(4):1964–70.
216. Bergougnoux, V., Caissard, J. C., Jullien, F. et al. 2007. Both the adaxial and abaxial epidermal layers of the rose petal emit volatile scent compounds. *Planta* 226(4):853–66.
217. Rowan, D. D. 2011. Volatile metabolites. *Metabolites* 1(1):41–63.
218. Gu, X. G., Zhang, Z. Z., Wan, X. C., Ning, J. M., Yao, C. C. and Shao, W. F. 2009. Simultaneous distillation extraction of some volatile flavor components from Pu-erh tea samples-comparison with steam distillation-liquid/liquid extraction and Soxhlet extraction. *Int. J. Anal. Chem.* 2009(1687–8760):276713.
219. Chaintreau, A. 2001. Simultaneous distillation–extraction: From birth to maturity—review. *Flavour & Frag. J.* 16(2):136–48.

220. Kumazawa, K. and Masuda, H. 1999. Identification of potent odorants in Japanese green tea (Sen-cha). *J. Agric. Food Chem.* 47(12):5169–72.
221. Kumazawa, K., Wada, Y. and Masuda, H. 2006. Characterization of epoxydecenal isomers as potent odorants in black tea (Dimbula) infusion. *J. Agric. Food Chem.* 54(13):4795–801.
222. Kawakami, M. 1997. Comparison of Extraction Techniques for Characterizing Tea Aroma and Analysis of Tea by GC-FTIR-MS. In: Linskens, H. F., Jackson, J. F., (Eds) *Modern Methods of Plant Analysis, vol. 19*, Plant Volatile Analysis, pp. 211–29.
223. Wang, D. M., Yoshimura, T, Kubota, K and Kobayashi, A. 2000. Analysis of glycosidically bound aroma precursors in tea leaves. 1. Qualitative and quantitative analyses of glycosides with aglycons as aroma compounds. *J. Agric. Food Chem.* 48(11):5411–18.
224. Yang, Z., Tomomi, K., Aya, T., Hironori, S., Akio, M. and Naoharu, W. 2009. Analysis of coumarin and its glycosidically bound precursor in Japanese green tea having sweet-herbaceous odour. *Food Chem.* 114(1):289–94.
225. Engel, W., Bahr, W. and Schieberle, P. 1999. Solvent assisted flavour evaporation–a new and versatile technique for the careful and direct isolation of aroma compounds from complex food matrices. *Eur. Food Res. Technol.* 209(3):237–41.
226. Xu, Yan, Fan, Wenlai and Qian, M. C. 2007. Characterization of aroma compounds in apple cider using solvent-assisted flavor evaporation and headspace solid-phase microextraction. *J. Agric. Food Chem.* 55(8):3051–7.
227. Vichi, S., Guadayol, J. M., Caixach, J., Lópeztamame, E. S. and Buxaderas, S. 2007. Comparative study of different extraction techniques for the analysis of virgin olive oil aroma. *Food Chem.* 105(3):1171–8.
228. Bonino, M., Schellino, R., Rizzi, C., Aigotti, R., Delfini, C. and Baiocchi, C. 2003. Aroma compounds of an Italian wine (Ruché) by HS–SPME analysis coupled with GC–ITMS. *Food Chem.* 80(1):125–33.
229. Poinot, P., Grua-Priol, J., Arvisenet, G., et al. 2007. Optimisation of HS-SPME to study representativeness of partially baked bread odorant extracts. *Food Res. Int.* 40(9):1170–84.
230. Cheong, K. W., Tan, C. P., Mirhosseini, H., Nsa, H., Osman, A. and Basri, M. 2010. Equilibrium headspace analysis of volatile flavor compounds extracted from soursop (*Annona muricata*) using solid-phase microextraction. *Food Res. Int.* 43(5):1267–76.
231. Wei, Q., Wang, H., Lv, Z., et al. 2013. Search for potential molecular indices for the fermentation progress of soy sauce through dynamic changes of volatile compounds. *Food Res. Int.* 53(1):189–94.
232. Xin-Guo, S. U., Jiang, Y. M., Wang, X. H., Duan, J., Sui-Qing, M. I. and Wang, N. S. 2005. Application of solid phase microextraction on the oolong tea (*Camellia sinensis* cv. Fenghuangdancong) aroma analysis. *Food Sci.* 11(3):308–11.
233. Mo, X., Xu, Y. and Fan, W. 2010. Characterization of aroma compounds in Chinese rice wine Qu by solvent-assisted flavor evaporation and headspace solid-phase microextraction. *J. Agric. Food Chem.* 58(4):2462–9.
234. Jumtee, K., Komura, H., Bamba, T. and Fukusaki, E. 2011. Predication of Japanese green tea (Sen-cha) ranking by volatile profiling using gas chromatography mass spectrometry and multivariate analysis. *J. Biosci. Bioeng.* 112(3):252–5.
235. Tontul, I., Torun, M., Dincer, C., et al. 2013. Comparative study on volatile compounds in Turkish green tea powder: Impact of tea clone, shading level and shooting period. *Food Res. Int.* 53(2):744–50.
236. Dong, F., Yang, Z., Baldermann, S., et al. 2012. Characterization of L-phenylalanine metabolism to acetophenone and 1-phenylethanol in the flowers of *Camellia sinensis* using stable isotope labeling. *J. Plant Physiol.* 169(3):217–25.
237. Schuh, C. and Schieberle, P. 2006. Characterization of the key aroma compounds in the beverage prepared from Darjeeling black tea: Quantitative differences between tea leaves and infusion. *J. Agric. Food Chem.* 54(3):916–24.
238. Tholl, D. and Röse, U. S. R. 2006. Detection and identification of floral scent compounds. In: Dudareva N., Pichersky E., (Eds.), *Biology of Floral Scent*, Boca Raton, FL: CRC Press/Taylor & Francis Group, pp. 3–25.

239. And, K. K. and Masuda, H. 2002. Identification of potent odorants in different green tea varieties using flavor dilution technique. *J. Agric. Food Chem.* 50(20):5660–3.
240. Guth, H. and Grosch, W. 1993. Identification of potent odourants in static headspace samples of green and black tea powders on the basis of aroma extract dilution analysis (AEDA). *Flavour Frag. J.* 8(4):173–8.
241. Dutta, R., Hines, E. L., Gardner, J. W., Kashwan, K. R. and Bhuyan, M. 2003. Tea quality prediction using a tin oxide-based electronic nose: An artificial intelligence approach. *Sens. Actuators B Chem.* 94(2):228–37.
242. Bhattacharyya, N., Seth, S., Tudu, B., et al. 2007. Monitoring of black tea fermentation process using electronic nose. *J. Food Eng.* 80(4):1146–56.
243. Yu, H., Wang, J., Zhang, H., Yu, Y. and Yao, C. 2008. Identification of green tea grade using different feature of response signal from E-nose sensors. *Sens. Actuators B Chem.* 128(2):455–61.
244. Yang, Z. Y., Fang, D., Shimizu, K., et al. 2009. Identification of coumarin-enriched Japanese green teas and their particular flavor using electronic nose. *J. Food Eng.* 92(3):312–16.
245. Kovács, A., Simonsarkadi, L. and Ganzler, K. 1999. Determination of biogenic amines by capillary electrophoresis. *J. Chromatogr. A* 836(2):305–13.
246. Gardener, M. C. and Gillman, M. P. 2001. Analyzing variability in nectar amino acids: Composition is less variable than concentration. *J. Chem. Ecol.* 27(12):2545–58.
247. Armenta, J. M., Cortes, D. F., Pisciotta, J. M., et al. 2010. A sensitive and rapid method for amino acid quantitation in malaria biological samples using AccQ•Tag UPLC-ESI-MS/MS with multiple reaction monitoring. *Anal. Chem.* 82(2):548–58.
248. Hall, M. N., Robertson, A. and Scotter, C. N. G. 1988. Near-infrared reflectance prediction of quality, theaflavin content and moisture content of black tea. *Food Chem.* 27(1):61–75.
249. Sinija, V. R. and Mishra, H. N. 2009. FT-NIR spectroscopy for caffeine estimation in instant green tea powder and granules. *LWT - Food Sci. and Tech.* 42(5):998–1002.
250. Friedman, M., Levin, C. E., Choi, S. H., Kozukue, E. and Kozukue, N. 2006. HPLC analysis of catechins, theaflavins, and alkaloids in commercial teas and green tea dietary supplements: Comparison of water and 80% ethanol/water extracts. *J. Food Sci.* 71(6):C328–C337.
251. Matsui, Y., Kobayashi, K., Masuda, H., et al. 2009. Quantitative analysis of saponins in a tea-leaf extract and their antihypercholesterolemic activity. *Biosci. Biotechnol. Biochem.* 73(7):1513–19.
252. Verza, S. G., Silveira, F., Cibulski, S., et al. 2012. Immunoadjuvant activity, toxicity assays, and determination by UPLC/Q-TOF-MS of triterpenic saponins from *Chenopodium quinoa* seeds. *J. Agric. Food Chem.* 60(12):3113–8.
253. Qian, Z. M., Lu, J., Gao, Q. P. and Li, S. P. 2009. Rapid method for simultaneous determination of flavonoid, saponins and polyacetylenes in Folium Ginseng and Radix Ginseng by pressurized liquid extraction and high-performance liquid chromatography coupled with diode array detection and mass spectrometry. *J. Chromatogr. A.* 1216(18):3825–30.
254. Kaličanin, B. and Velimirović, D. 2013. The content of lead in herbal drugs and tea samples. *Cent. Eur. J. Biol.* 8(2):178–85.
255. Al, O. S. S. 2003. Heavy metal contents in tea and herb leaves. *Pak. J. Bio. Sci.*, 6:208–12.
256. Milani, R. F., Morgano, M. A., Saron, E. S., Silva, F. F. and Cadore, S. 2015. Evaluation of direct analysis for trace elements in tea and herbal beverages by ICP-MS. *J. Braz. Chem. Soc.* 26(6):1211–17.
257. Mahani, M. K. and Maragheh, M. G. 2011. Simultaneous determination of sodium, potassium, manganese and bromine in tea by standard addition neutron activation analysis. *Food Anal. Methods.* 4(1):73–6.
258. Szymczycha-Madeja, A., Welna, M. and Pohl, P. 2012. Elemental analysis of teas and their infusions by spectrometric methods. *TrAC, Trends Anal. Chem.* 35(5):165–81.
259. Sinija, V. R., Mishra, H. N. and Bal, S. 2007. Process technology for production of soluble tea powder. *J. Food Eng.* 82(3):276–83.

260. Someswararao, C. and Srivastav, P. P. 2012. A novel technology for production of instant tea powder from the existing black tea manufacturing process. *Innov. Food Sci. Emerging Technol.* 16(39):143–7.
261. Quan, V. V., Golding, J. B., Nguyen, M. H. and Roach, P. D. 2013. Preparation of decaffeinated and high caffeine powders from green tea. *Powder Technol.* 233(2):169–75.
262. Nadeem, H. Ş., Torun, M. and Özdemir, F. 2011. Spray drying of the mountain tea (*Sideritis stricta*) water extract by using different hydrocolloid carriers. *LWT - Food Sci. Technol.* 44(7):1626–35.
263. Xia, T., Shi, S. and Wan, X. 2006. Impact of ultrasonic-assisted extraction on the chemical and sensory quality of tea infusion. *J. Food Eng.* 74(4):557–60.
264. Alasalvar, C., Pelvan, E., Ozdemir, K. S., et al. 2013. Compositional, nutritional, and functional characteristics of instant teas produced from low- and high-quality black teas. *J. Agric. Food Chem.* 61(31):7529–36.
265. Ekborgott, K. H., Taylor, A. and Armstrong, D. W. 1997. Varietal differences in the total and enantiomeric composition of theanine in tea. *J. Agric. Food Chem.* 45(2):353–63.
266. Juneja, L. R., Chu, D. C., Okubo, T., Nagato, Y. and Yokogoshi, H. 1999. L-theanine — a unique amino acid of green tea and its relaxation effect in humans. *Trends Food Sci. Technol.* 10(6–7):199–204.
267. Scharbert, S., Holzmann, N. and Hofmann, T. 2004. Identification of the astringent taste compounds in black tea infusions by combining instrumental analysis and human bioresponse. *J. Agric. Food Chem.* 52(11):3498–508.
268. Scharbert, S. and Hofmann, T. 2005. Molecular definition of black tea taste by means of quantitative studies, taste reconstitution, and omission experiments. *J. Agric Food Chem.* 53(13):5377–84.
269. Kaneko, S., Kumazawa, K., Masuda, H., Henze, A. and Hofmann, T. 2006. Molecular and sensory studies on the umami taste of Japanese green tea. *J. Agric. Food Chem.* 54(7):2688–94.
270. Yu, P., Yeo, A. S., Low, M. Y. and Zhou, W. 2014. Identifying key non-volatile compounds in ready-to-drink green tea and their impact on taste profile. *Food Chem.* 155(2):9–16.

Chapter 4

The potential role for tea in combating chronic diseases

Chung S. Yang, Rutgers University, USA

1 Introduction

Tea, made from the leaves of the plant *Camellia sinensis*, has been used for medicinal purposes since ancient days and is now a popular beverage. During the past 30 years, tea, especially green tea polyphenols, has been studied for its potential beneficial health effects. These include the prevention of cancer, diabetes, cardiovascular diseases (CVDs) and neurodegenerative diseases; reduction of body weight and alleviation of metabolic syndome (MetS) (reviewed in Yang and Hong, 2013; Yang et al. 2009, 2011a, 2016). Most of these beneficial effects are believed to be due to the presence of polyphenols in green tea, although caffeine also contributes to some of the effects. A unique amino acid, theanine (γ-ethylamino-L-glutamic acid), has been shown to have neuroprotective effects.

Cancer and CVDs are the two most common diseases and the top two leading causes of death in many countries. Overweight, obesity and diabetes are emerging as major health issues, and the closely related MetS also predisposes individuals to CVDs. Neurodegenerative diseases are becoming major social concerns because of the ageing of the world's population. If tea could prevent or delay the development of these diseases, the public health implications would be substantial. Because of this, there is immense

http://dx.doi.org/10.19103/AS.2017.0036.18

scientific and public interest on this topic. A literature search on *PubMed* using the key words 'Tea and Health' yielded 3831 publications, 'Tea and cancer' – 3657 publications, 'Tea and diabetes' – 640 publications, 'Tea and heart disease' – 142 publications and 'Tea and weight control' – 290 publications. Unfortunately, some of the beneficial effects found in the laboratory may have been over-extrapolated to human health and propagated in the news media, popular magazines and even review articles.

This chapter reviews these topics, including possible human relevance of the published results. Selected examples are used to illustrate the beneficial health effects and possible mechanisms involved. Information from recently published meta-analyses and systematic reviews are used to help assess the relative strengths of the existing data. Suggestions for future research are made. The author believes this article will enhance our understanding of the health effects of tea consumption.

2 Chemical properties, bioavailability and biotransformation of tea constituents

2.1 Tea constituents and their properties

Depending on the post-harvest processing of the tea leaves, the common types of tea consumed are green tea, black tea and oolong tea, with black tea accounting for more than 78% of the world tea production. In the manufacturing of green tea, the tea leaves are heated or steamed, rolled and dried, which inactivates the enzymes with minimum oxidation of the constituents. The drying of the tea leaves helps to stabilize the tea constituents upon storage. Green tea possesses characteristic polyphenolic compounds known as catechins, which include (–)-epigallocatechin-3-gallate (EGCG), (–)-epigallocatechin (EGC), (–)-epicatechin-3-gallate (ECG) and (–)-epicatechin (EC). The structures of catechins together with theaflavins from black tea are shown in Fig. 1. Tea leaves also contain lower quantities of other polyphenols such as quercetin, kaempferol and myricetin, as well as alkaloids such as caffeine and theobromine. A typical brewed green tea beverage (e.g. 2.5 g tea leaves in 250 ml of hot water) contains 240–320 mg of catechins, of which 60–65% is EGCG, with 20–50 mg of caffeine (Balentine et al., 1997; Sang et al., 2011).

In the manufacturing of black tea, the tea leaves are withered, crushed and allowed to undergo enzyme-mediated oxidation in a process commonly referred to as 'fermentation'. During this process, most of the catechins are oxidized, dimerized and polymerized to form theaflavins and thearubigins (Balentine et al., 1997; Sang et al., 2011). Theaflavins are produced from the dimerization of two catechin molecules. Theaflavins exist in four forms (theaflavin, theaflavin-3 gallate, theaflavin-3'-gallate and theaflavin-3,3'-digallate) and contribute to the orange colour and characteristic taste of black tea. Thearubigins are heterogeneous polymers of tea catechins with reddish brown colour, but the structures are poorly understood. In brewed black tea, catechins, caffeine, theaflavins and thearubigins each account for 3–10%, 3–6%, 2–6% and greater than 20% of the dry weight, respectively. Oolong tea is manufactured by crushing only the rims of the leaves and limiting fermentation to a short period to produce specific flavour and taste of the tea. Oolong tea contains catechins, theaflavins and thearubigins as well as some characteristic components: epigallocatechin esters, dimeric catechins (such as theasinensins) and dimeric proanthocyanidins (Sang et al., 2011).

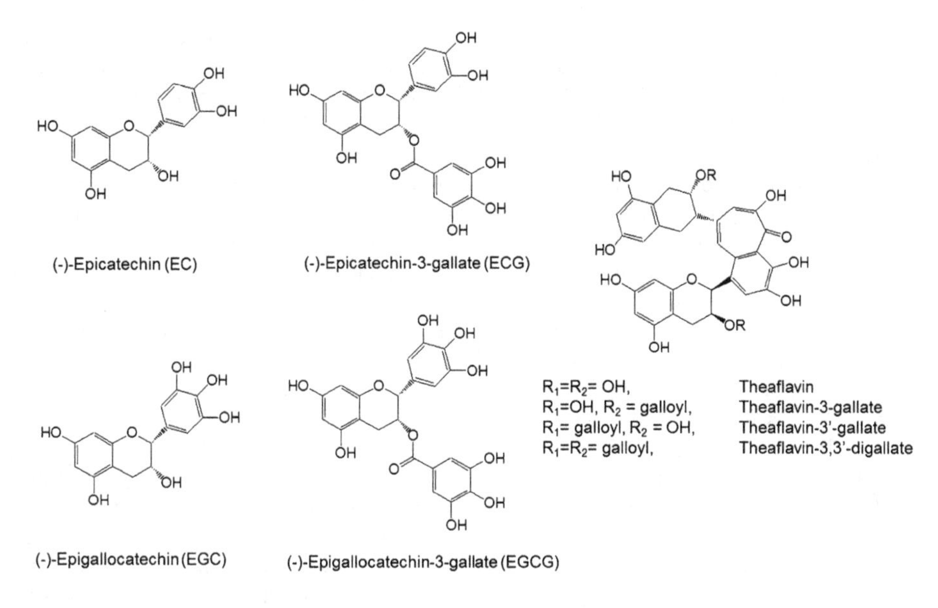

Figure 1 Structures of catechins and theaflavins.

Tea catechins are strong antioxidants, scavenging free radicals and also preventing the formation of reactive oxygen species (ROS) by chelating metal ions (reviewed in Sang et al., (2011)). *In vivo*, EGCG and other catechins can serve as antioxidants, but they may also cause the formation of ROS in the mitochondria under certain conditions (Li et al., (2010), Tao et al. (2014)). The ROS may activate nuclear factor erythroid 2-related factor 2 (Nrf2)-mediated antioxidant and other cytoprotective enzymes (Shen et al., 2005; Wang et al., 2015; James et al., 2015). EGCG is also known to undergo superoxide-catalysed auto-oxidation *in vitro* to produce ROS that can induce cell death (Hou et al., 2005). Nevertheless, such auto-oxidation of EGCG is unlikely to occur in internal organs, because of the lower oxygen partial pressure (than in solution *in vitro*) and the presence of antioxidant enzymes in animal tissues (Hou et al., 2005). Therefore, results on EGCG obtained from cell culture studies need to be interpreted with caution.

An important biochemical property of catechins is their H-bonding, via their phenolic groups, to proteins, lipids and nucleic acids. The multiple H-bond formation provides high-affinity binding to these biomolecules. As will be discussed subsequently, the binding of EGCG to many proteins has been proposed to be a key mechanism for its anti-cancer activities. Black tea polyphenols, with more phenolic groups, may bind to biomolecules with even higher affinity than EGCG.

2.2 Bioavailability and biotransformation

According to Lipinski's Rule of 5 (Lipinski et al., 2001), compounds that have 5 or more hydrogen bond donors, 10 or more hydrogen bond acceptors, a molecular weight greater than 500 and Log P greater than 5 are usually poorly absorbed following oral administration. This is due to their large actual size (high molecular weight), large apparent size (due to the formation of a large hydration shell) and high polarity (Lipinski et al., 2001). The bioavailabilities of tea polyphenols follow this prediction (reviewed in Yang et al. (2008a), Chow and Hakim (2011)). Both human and animal studies have shown that the bioavailability of EC (molecular weight 290 and 5 phenolic groups) is higher than that of EGCG (molecular weight 458 and 8 phenolic groups). In rats, following intragastric (i.g.) administration of decaffeinated green tea (200 mg/kg), the absolute plasma bioavailabilities of EGCG, EGC and EC were 0.1, 14 and 31%, respectively. EGCG, EGC and EC in plasma had elimination half-lives of 165, 66 and 67 min, respectively (Yang et al., 2008b). However, the absolute plasma bioavailability of EGCG in mice following i.g. administration of EGCG (75 mg/kg) was much higher at 26.5%, with more than 50% of EGCG present as glucuronide conjugates. Levels of EGCG in the small intestine and colon were 20.6 and 3.6 ng/g, respectively (Yang et al., 2008b). In humans, following oral administration of the equivalent of two or three cups of green tea, the peak plasma levels of tea catechins (including the conjugated forms) were usually 0.2–0.3 μM. With high pharmacological oral doses of EGCG, peak plasma concentrations of 2–9 μM and 7.5 μM were reported in mice and humans, respectively (Yang et al., 2008a). Conversely, theaflavin and theaflavin-3,3′-digallate (molecular weights of 564 and 868 and containing 9 and 14 phenolic groups, respectively) were reported to have extremely low bioavailability when administered orally (Mulder et al., 2001). However, more studies are needed in this area.

EGCG and other tea catechins undergo extensive biotransformation (reviewed in Yang et al. (2008a)). Because of the catechol structure, EGCG and other catechins are readily methylated by catechol-*O*-methyltransferase. EGC is also readily methylated to form 4′-*O*-methyl-EGC. This metabolite as well as 4″-*O*-methyl-EGCG and 4′,4″-dimethyl-EGCG have been detected in human and animal plasma and urine samples after the ingestion of tea. In addition to methylation, catechins are also glucuronidated by UDP-glucuronosyltransferases and sulphated by sulphotransferases. Multiple methylation and conjugation reactions can occur on the same molecule. For example, we have observed methyl-EGCG-glucuronide, EGCG-glucuronide-sulphate, dimethyl-EGCG-diglucuronide and methyl-EGCG-glucuronide-sulphate as urinary metabolites in mice (Sang et al., 2011).

Active efflux has been shown to limit the bioavailability of many polyphenolic compounds. The multidrug resistance-associated protein 2 (MRP2), located on the apical surface of the intestine and liver, mediates the transport of some polyphenolic compounds to the lumen and bile, respectively (Jemnitz et al., 2010). Therefore, EGCG and its metabolites are predominantly effluxed from the enterocytes into the intestinal lumen, or effluxed from the liver to the bile and excreted in the faeces, with little or none of these compounds in the urine of humans and rats. However, low levels of urinary EGCG metabolites (in the conjugated forms) can be detected in mice (Sang et al., 2011). Tea catechins can be degraded in the intestinal tract by microorganisms. We have observed the formation of ring fission metabolites 5-(3′,4′,5′-trihydroxyphenyl)-γ-valerolactone (M4), 5-(3′,4′-dihydroxyphenyl)-γ-valerolactone (M6) and 5-(3′,5′-dihydroxyphenyl)-γ-valerolactone (M6′) in human urine and plasma samples several hours after the ingestion of tea (Li et al., 2000; Lee et al., 2002). These compounds can undergo further degradation to phenylacetic and phenylpropionic acids.

Several investigators have reported the pharmacokinetics of tea polyphenols in human volunteers. For example, we found that in the oral administration of 20 mg green tea solids per kg body weight, it took 1.4–1.6 h for the catechins to reach peak plasma concentrations (Lee et al., 2002). This resulted in maximal plasma concentrations of 77.9, 223 and 124 ng/mL, respectively, for EGCG, EGC and EC. EGCG, EGC and EC had terminal half-lives of

3.4, 1.7 and 2 h, respectively. Plasma EGC and EC were present mainly in the conjugated forms, whereas 77% of the EGCG was in the free form. Methylated EGCG and EGC were also present in human plasma (Yang et al., 2008a; Lee et al., 2002). Chow et al. (2003) demonstrated that, following four weeks of oral administration of EGCG (800 mg, once daily), there was an increase in the systemic bioavailability, but the molecular basis for this observation remains to be investigated.

In contrast to the limited bioavailabilities of catechins, Caffeine has bioavailability close to 100% and is mainly metabolized by cytochromes P450 1A2 to dimethylxanthines and theophylline (Arnaud, 2011). Theanine is also almost 100% bioavailable and is metabolized to ethylamine and glutamic acid (Scheid et al., 2012; van der Pijl et al., 2010). Both caffeine and theanine readily cross the blood–brain barrier and are neurologically active.

3 Tea and cancer prevention

3.1 Inhibition of carcinogenesis in animal models

Tea and its major constituents have been demonstrated to inhibit tumorigenesis in many animal models for different organ sites, including the lung, oral cavity, oesophagus, stomach, small intestine, colon, skin, liver, pancreas, bladder, prostate and mammary glands (reviewed in Yang et al. (2009, 2011a)). Most of the studies were conducted with green tea, green tea polyphenol preparations or pure EGCG, administered through the drinking water or diet. Some of the examples are discussed as follows.

At least 20 studies have demonstrated the inhibitory effect of tea or tea preparations on lung tumorigenesis (reviewed in Yang et al. (2011a)). Most of the experiments were conducted in tobacco carcinogen-treated or transgenic mouse models and a few studies in rat and hamster models. Inhibitory activities have been demonstrated when green tea preparations were administered during the initiation, promotion or progression stages of carcinogenesis. Treatment with green or black tea extracts for 60 weeks also inhibited the spontaneous formation of lung tumours as well as rhabdomyosarcomas in A/J mice. These results demonstrate the broad activities of tea preparations in the inhibition of lung neoplasia at different stages of carcinogenesis.

Inhibitory effects of tea against tumorigenesis in the digestive tract, including the oral cavity, oesophagus, stomach, small intestine and colon, have been shown in more than 30 studies (Yang et al., 2011a). For example, tea preparations were shown to inhibit chemically induced oral carcinogenesis in a hamster model and oesophageal carcinogenesis in a rat model. EGCG also inhibited tumorigenesis in rat stomach and forestomach induced by *N*-methyl-*N*-nitro-*N*-nitrosoguanidine. The inhibitory effects of tea and tea polyphenols on intestinal tumorigenesis in mice have been consistently observed in different laboratories. The effects of tea preparations on colon tumorigenesis in rats, however, have not been consistent (Yang et al., 2011a). Our recent animal studies showed that in AOM-treated rats, administration of Polyphenon E (PPE; a standardized tea polyphenol preparation that contains 65% EGCG, 25% other catechins and 0.6% caffeine), at a dose of 0.24% in the diet, decreased ACF formation, adenocarcinoma incidence and adenocarcinoma multiplicity (Yang et al., 2011a).

As reviewed previously (Yang et al., 2009, 2011a), most of the reported studies have demonstrated the cancer-preventive activities of tea catechins against carcinogenesis at different organ sites and the evidence is strong. However, there are also some studies that did not observe cancer-preventive effects. The reasons for the discrepancies may be complex. One possible reason is the dose and bioavailability of the tea catechins used. For example, in studying the prevention of mammary carcinogenesis, the concentrations of EGCG in the mammary tissues may be too low to be effective in some of the experiments (Yang et al., 2011a). Although EGCG and other catechins are thought to be the major cancer-preventive agents in tea, effective inhibition of carcinogenesis by caffeine in the lung and skin, but not in the colon, has been demonstrated (Yang et al., 2011a).

3.2 Tea consumption and cancer risk in humans

3.2.1 Observational epidemiological studies

In contrast to the strong evidence for the cancer-preventive activities of tea constituents in animal models, results from epidemiological studies have not been consistent in demonstrating the cancer-preventive effect of tea consumption in humans. A comprehensive review by Yuan et al. (2011) concluded that consumption of green tea was frequently associated with a reduced risk of upper gastrointestinal tract cancer, after adjusting for confounding factors and limited data supported its protective effect of lung and hepatocellular carcinogenesis. However, intake of black tea in general was not associated with a lower risk of cancer (Yuan et al., 2011). We agree with this general conclusion, and would like to add the following discussions.

In a meta-analysis of 25 epidemiological studies on tea consumption and colorectal cancer published in 2006 (Sun et al., 2006), no association was found between black tea and colorectal cancer, and the results on green tea were mixed. Subsequently, a prospective study on women in Shanghai found a reduced risk of colorectal cancer in green tea drinkers (Yang et al., 2007). On the other hand, a cohort study in Singapore found that green tea consumption had a statistically non-significant increased risk for advanced stage colon cancer only in men (Sun et al., 2007). The observed adverse effect in men may be related to smoking or the pro-inflammatory activities of tea catechins as were observed in animal models (Guan et al., 2012).

In a meta-analysis on the relation between green tea consumption and breast cancer published in 2010, an inverse association was found in the four case–control studies but not in the three cohort studies (Ogunleye et al., 2010). Two additional cohort studies in Japan and China also did not find an association. These results are consistent with the rather weak evidence from animal studies on the prevention of mammary cancer by tea. Similarly, an inverse association between green tea consumption and prostate cancer was found in two case–control studies, but not in four prospective cohort studies (Yuan et al., 2011).

Smoking appears to be a strong interfering factor in studies on digestive tract cancers. For example, in a case–control study on the effect of green tea consumption on oesophageal cancer in Shanghai, a protective effect was only observed in women, who were mostly non-smokers (Gao et al., 1994). Similarly, in the Shanghai Men's Health Study, green tea drinking was found to reduce the risk of colorectal cancer among non-smoking men, but not for men in general (Yang et al., 2011b). In a recent large-scale, population-based case–control study in urban Shanghai, regular green tea drinking was associated with a significant reduction of pancreatic cancer risk in women – who were mostly non-smokers, but not in men – who were mostly smokers and former smokers (Wang et al., 2012). A recent systematic review of epidemiological studies in Japan on green tea consumption and gastric cancer indicated no overall preventive effect of green tea in cohort studies. However, a small consistent risk reduction was found in women (mostly non-smokers), and the result was statistically significant in the pooled data of six cohort studies (Sasazuki et al., 2012).

3.2.2 Intervention studies

Many intervention trials have been conducted with green tea, and some have found a beneficial effect of green tea polyphenols in preventing the development or progression of cancer. For example, in a double-blinded, Phase II trial in Italy, 30 men with high-grade prostate intraepithelial neoplasia (PIN) were given 300 mg of green tea catechins twice daily for 12 months (Bettuzzi et al., 2006). Only one patient developed prostate cancer, whereas nine of the 30 patients with high-grade PIN in the placebo group developed prostate cancer (highly statistically significant). However, in a recent trial in Florida with a similar design using PPE (containing 400 mg of EGCG) in 97 men with high-grade PIN and/or atypical small acinar proliferation (ASAP), supplementation for 6–12 months showed no differences in the number of observed prostate cancer cases between the treatment group (n = 49) and the placebo group (n = 48) (Kumar et al., 2015). Nevertheless, there was a decrease in the cumulative rate of prostate cancer plus ASAP in the treatment group, in subjects without ASAP at baseline. A decrease in serum prostate–specific antigen was also observed in the PPE-supplemented group (Kumar et al., 2015).

An earlier randomized controlled trial (RCT) on oral cancer prevention in China, with a mixed tea product (3 g/day administered orally or topically) in patients with oral mucosa leukoplakia for six months, showed significant decrease in the number and total volume of proliferation index and silver-stained nucleoli organizer regions (Li et al., 1999). Nevertheless, a later Phase II RCT in the United States with green tea extract (GTE; 500, 750 or 1000 mg/m^2, two times daily) for 12 weeks, to patients with oral pre-malignant lesions (n = 28), showed possible beneficial effects in lessening oral pre-malignant lesions, in part through reducing angiogenic stimulus (stroma vascular endothelial growth factor), but it was not statistically significant (Tsao et al., 2009). Some recent intervention studies on breast cancer and oesophageal adenocarcinoma were limited to bioavailability and some biomarker studies (Crew et al., 2012; Joe et al., 2015). At present, the earlier optimistic expectation of cancer-preventive activity by tea polyphenols, based on laboratory results, has not materialized in RCTs. More than 20 human trials with green tea polyphenol preparations are ongoing in the United States, China and Japan (NIH clinical trials website[1]). Some of these studies may yield clear conclusions concerning cancer-preventive activities of green tea polyphenols.

3.3 Mechanistic considerations

Many mechanisms have been proposed for cancer prevention by tea constituents, and this subject has been reviewed (Yang et al., 2009, 2011a). ROS have been shown to play key roles in carcinogenesis; the antioxidant actions of tea catechins could be an important mechanism for cancer prevention. Another possible mechanism is through the binding of EGCG to target proteins, leading to the inhibition of metabolic or signal transduction pathways. As reviewed previously, the 67-kDa laminin receptor, Bcl-2 proteins, vimentin, peptidyl prolyl *cis/trans* isomerase (Pin 1) and other proteins have been proposed as targets for EGCG (Yang et al., 2009, 2011a). It is reasonable to assume that the high-affinity

[1] http://clinicaltrials.gov search on 30 August 2016.

binding proteins reported in the literature could serve as initial targets, but this point remains to be substantiated in animal models. Some of the proposed mechanisms based on studies in cell lines, however, may not be relevant to cancer prevention.

Apparently, mechanisms derived from cancer prevention studies in animal models are likely to be more relevant. These include the induction of apoptosis in different animal models, inhibition of the phosphorylation of c-Jun and Erk1/2 in lung tumorigenesis model, suppression of phospho-Akt and nuclear β-catenin levels in colon cancer models, inhibition of the IGF/IGF-1R axis in colon, prostate and other cancer models and suppression of vascular endothelial growth factor–dependent angiogenesis in lung and prostate cancer models (Yang et al., 2011a). It is still unclear whether these molecules are direct targets for EGCG or downstream events of the primary action. Based on the limited human data, actions of tea polyphenols in reducing oxidative stress and enhancing the elimination of carcinogens may be considered as important mechanisms.

4 Reduction of body weight, alleviation of metabolic syndrome and prevention of diabetes

Overweight, obesity and type 2 diabetes are emerging as major health issues in many countries. MetS is a complex of symptoms that include enlarged waist circumference and two or more of the following: elevated serum triglyceride, dysglycemia, high blood pressure and reduced high-density lipoprotein-associated cholesterol (Ford, 2005). The possible beneficial effects of tea consumption on body weight reduction and MetS alleviation could have huge public health implications.

4.1 Studies in animal models

The effects of tea, tea polyphenols and EGCG on body weight and MetS have been studied extensively in animal models (reviewed in Yang and Hong (2013), Huang et al. (2014), Yang et al. (2016)). Most of the studies showed that consumption of GTE or EGCG significantly reduced the gaining of body weight and/or adipose tissue weight, lowered blood glucose or insulin levels and increased insulin sensitivity or glucose tolerance. Most of these studies used high-fat diets or genetically induced obese/diabetic rodent models. For example, in mice fed with a high-fat (60% of the calories) diet, we found that dietary EGCG treatment (0.32% in diet) for 16 weeks significantly reduced body weight gain, body fat and visceral fat weight (Bose et al., 2008). These results were also reproduced in a second study using a high-fat/Western-style diet (Chen et al., 2011). EGCG treatment also attenuated insulin resistance, plasma cholesterol and monocyte chemoattractant protein levels in mice on high-fat and high-fat/Western diets (Bose et al., 2008; Chen et al., 2011). Similar results were also observed in several recent studies (Okuda et al., 2014; Byun et al., 2014; Ortsater et al., 2012; Lee et al., 2015). Green tea polyphenols also alleviate MetS in other animal models. For example, in insulin-resistant rats, treatment with green tea polyphenols significantly decreased blood glucose, insulin, triglycerides, total cholesterol, low-density lipoprotein (LDL) cholesterol and free fatty acids (Qin et al., 2010). In insulin-resistant beagle dogs, oral administration of GTE (80 mg/kg daily, before the daily meal) for 12 weeks also markedly increased insulin sensitivity index (Serisier et al., 2008).

Diet-induced liver steatosis, which predisposes to liver cancer, is becoming a common disease, and its possible prevention by tea consumption warrants more investigation. We have shown that EGCG treatment reduces the incidence of hepatic steatosis, liver size (48% decrease), liver triglycerides (52% decrease), plasma alanine aminotransferase concentration (67% decrease) and liver pathology in mice fed with a high-fat diet (Bose et al., 2008, Chen et al., 2011). Tea catechins have been reported to also reduce hepatic steatosis and liver toxicity in rodents treated with ethanol, tamoxifen or endotoxins, or rodents with liver ischaemia/reperfusion injury (reviewed in Sae-tan et al. (2011)). These findings have potential for practical applications.

4.2 Studies in humans

4.2.1 Randomized controlled trials

The effects of tea consumption on body weight and biomarkers of MetS have been studied in many short-term RCTs during the past decade. Systematic reviews and meta-analysis covering more than 26 earlier RCTs indicated the beneficial effects of tea consumption in reducing body weight and alleviating MetS (Hursel et al., 2009; Phung et al., 2010). Most of these studies used green tea or GTE with caffeine, in studies for 8–12 weeks, on normal weight or overweight subjects. Some of the more recent RCTs also showed that daily consumption of 458–886 mg of green tea catechins by moderately overweight Chinese subjects for 90 days reduced body fat (Wang et al., 2010), and that daily intake of PPE capsules (containing 400 or 800 mg EGCG and lower amounts of other catechins, but small amounts of caffeine) for two months by postmenopausal women in the United States decreased blood levels of LDL cholesterol, glucose and insulin (Wu et al., 2012). In another study, GTE supplementation (379 mg per day) to obese patients for three months decreased body weight and waist circumference (Suliburska et al., 2012). Improvements in lipid profiles, including the decrease in levels of total cholesterol, LDL cholesterol and triglycerides, were also observed (Suliburska et al., 2012). Similarly, an RCT in patients with obesity-related hypertension showed that consumption of GTE (379 mg daily) for three months reduced fasting serum glucose and insulin levels (Bogdanski et al., 2012).

A recent metabolomics study with healthy male subjects demonstrated that GTE supplementation (1200 mg catechins and 240 mg caffeine daily) for seven days increased lipolysis, fat oxidation and citric acid cycle activity under resting conditions without enhancing adrenergic stimulation (Hodgson et al., 2013). The role of caffeine in these studies was inconsistent among the different studies. A meta-analysis of metabolic studies showed that both a catechin–caffeine mixture and caffeine alone dose-dependently stimulated daily energy expenditure, but only the catechin-caffeine combination significantly increased fat oxidation (Hursel et al., 2011).

In contrast to the above-described beneficial effects, two recent studies in English adults did not show such beneficial effects (Hursel et al., 2011; Mielgo-Ayuso et al., 2014). The first study was in obese Caucasian women, after an energy-restricted diet intervention used supplementation with EGCG (200 mg daily) for 12 weeks (Mielgo-Ayuso et al., 2014). The second study used daily supplementation with green tea (>560 mg EGCG) plus caffeine (0.28–0.45 mg) for 12 weeks (Janssens et al., 2015). The reasons for these contradictory results are not known. The relatively low dose of EGCG used in the first study and the different populations, with different physiological and dietary conditions, could also be contributing factors. An intriguing issue is the time factor, that is, the increase in energy expenditure with catechins may be observed in short-term (7–10 days) studies, but not in long-term studies (Janssens et al., 2015). This may be related to our previous observations in mice that the plasma levels of EGCG increased in the first several days of daily catechin administration, but the EGCG levels decreased gradually to ~13% of the maximal levels upon prolonging treatment (Kim et al., 2000). More research is needed concerning the time-dependent changes in bioavailability and health beneficial effects of catechins.

4.2.2 Epidemiological studies

Two epidemiological studies suggested the beneficial effects of green tea consumption on MetS (Chang et al., 2012; Vernarelli and Lambert, 2013). One study on elderly Taiwanese males in a rural community indicated that tea drinking, especially for individuals who drank 240 ml or more tea daily, was inversely associated with incidence of MetS (Chang et al., 2012). The second, a cross-sectional study of US adults, showed that intake of hot (brewed) tea, but not iced tea, was inversely associated with obesity and biomarkers of MetS and CVDs (Vernarelli and Lambert, 2013). These results are exciting and need confirmation from additional studies. On the other hand, two recent cross-sectional studies in Japan did not find a preventive effect of green tea consumption against MetS (Takami et al., 2013; Pham et al., 2014).

The lowering of body weight and alleviation of MetS by tea should lead to the reduction of type 2 diabetes. Such an association was found in some, but not all, human studies (reviewed in Yang and Hong (2013), Huang et al. (2014), Wang et al. (2014) and Sae-tan et al. (2011)). For example, a prospective cross-sectional study with US women aged 45 years and older showed that consumption of more than four cups of tea per day was associated with a 30% lower risk of developing type 2 diabetes, whereas the consumption of total flavonoids or flavonoid-rich foods was not connected to reduced risk (Song et al., 2005). A retrospective cohort study of 17 413 Japanese adults aged 40–65 years indicated that daily drinking of more than six cups of green tea (but not oolong or black tea) lowered the morbidity of diabetes by 33% (Iso et al., 2006). The effects of caffeine in these epidemiological studies are unclear. A meta-analysis based on seven studies (286 701 total participants) showed that individuals who drank three to four or more cups of tea per day had a lower risk of type 2 diabetes than those consuming no tea (Huxley et al., 2009).

4.3 Mechanistic considerations

There are many proposed mechanisms for the above-described actions of tea polyphenols, and they can be summarized into two major types of actions. One is the action of tea polyphenols in the gastrointestinal tract and the other is the action produced by tea polyphenols after systemic absorption in different organs. The combined effects would reduce body weight, alleviate MetS and reduce the risk of diabetes and CVDs.

4.3.1 Actions in the gastrointestinal tract

Ingestion of green tea polyphenols has been shown to increase faecal lipid and total nitrogen contents, suggesting that polyphenols can decrease digestion and absorption of lipids and proteins (reviewed in Yang et al. (2016)). For example, in mice fed a high-fat diet, EGCG dose-dependently decreased food digestibility and increased the faecal mass;

with a ^{13}C-triglycerides-enriched diet, EGCG supplementation increased ^{13}C levels in the faeces (Friedrich et al., 2012).

The possibility that tea may affect gut microbiome has been studied in mice. For example, green tea powder feeding affected gut microbiota and reduced the levels of body fat, hepatic triglyceride and hepatic cholesterol; the reduction was correlated with the amount of *Akkermansia* and/or the total amount of bacteria in the small intestine (Axling et al., 2012). The abundance of *Akkermansia muciniphila* has been shown previously to be increased in prebiotic-treated *ob/ob* mice, which had lower fat mass compared to the control *ob/ob* mice (Everard et al., 2011). Changing gut microbiota, for example, by the administration of *Saccharomyces boulardii*, has also been shown to reduce hepatic steatosis, low-grade inflammation and fat mass in obese and type 2 diabetic *db/db* mice (Everard et al., 2014). In humans, green tea consumption has been reported to increase the proportion of the *Bifidobacterium* species in faecal samples (Jin et al., 2012). Increase in intestinal *Bifidobacteria* by a prebiotic (oligofructose) has been shown to decrease biomarkers for diabetes in mice (Cani et al., 2007). These results suggest the possibility that tea may alleviate MetS by enriching the probiotic population in the intestine. However, a recent study in humans indicated that long-term green tea consumption did not change the gut microbiota (Janssens et al., 2016). More studies in this area with green and black tea preparations are needed. The possibility that the EGCG activates the gut–brain–liver axis, intestinal endocrine systems or intestinal receptors could be important and warrants investigation.

4.3.2 Actions in internal organs

Many studies showed that ingestion of tea catechins suppressed gluconeogenesis and lipogenesis and enhanced lipolysis in a coordinated manner (reviewed in Yang et al. (2016)). These results suggest that the actions of tea catechins are mediated by energy-sensing molecules, possibly AMP (adenosine monophosphate)-activated protein kinase (AMPK). In response to falling energy status, AMPK is activated to inhibit energy-consuming processes and promote catabolism to produce adenosine triphosphate (ATP) (Long and Zierath, 2006; Hardie et al., 2012; Hardie, 2015). In addition to maintaining cellular energy homeostasis, AMPK also responds to different hormone signals to maintain whole body energy balance (Hardie, 2015). We propose that the activation of AMPK is the main mechanism for EGCG and other catechins to influence energy metabolism and to alleviate MetS. The activation of AMPK by EGCG and green tea polyphenols has been demonstrated *in vivo* and *in vitro* (Murase et al., 2009; Banerjee et al., 2012; Zhou et al., 2014; Collins et al., 2007; Serrano et al., 2013). There are also reports indicating that AMPK was activated in adipose tissues and skeletal muscle by black tea, oolong tea and Pu-erh teas (Yamashita et al., 2014; Yamashita et al., 2012). The detailed mechanism by which EGCG activates AMPK is still unclear, although the involvement of ROS has been suggested based on studies *in vitro* (Collins et al., 2007). The situation *in vivo* could be different due to the lack of auto-oxidation of EGCG. EGCG has been reported to inhibit mitochondrial oxidative phosphorylation to decrease ATP levels (Valenti et al., 2013). Another possibility is that EGCG may serve as an uncoupler of oxidative phosphorylation. Either action could increase the AMP (ADP) to ATP ratios and activate AMPK.

The downregulation of the two key enzymes in gluconeogenesis, PEPCK and G-6-Pase, and associated decrease in glucose production in the liver by EGCG has been shown to be mediated by AMPK activation (Collins et al., 2007). The activated form, p-AMPK, is also known to phosphorylate and inactivate ACC, the rate-limiting enzyme of fatty acid synthesis. The resulting lowered levels of malonyl-CoA can activate CPT-1, which facilitates long-chain fatty acyl CoA transport into mitochondria for β-oxidation (Long and Zierath, 2006). The possible role of AMPK in mediating the actions of tea constituents in affecting other genes and proteins to increase catabolism and decrease anabolism has been reviewed (Yang et al., 2016).

4.3.3 Overall mechanistic considerations

We proposed the hypothesis that most of these beneficial effects can be explained by the decreased absorption of macronutrients and the systemic effects of tea catechins after absorption in metabolic regulation, which are mediated mostly by the activation of AMPK. We further hypothesize that the relative importance of these two types of action depends on the types and amounts of tea consumed as well as the dietary conditions. For example, with black tea, the decrease in nutrient absorption, especially with a high-fat diet, may play a more important role than its systemic effects, because of the low bioavailability of theaflavins and thearubigins. Even though our 'AMPK hypothesis' proposes that AMPK plays a major role in mediating the actions of EGCG on gluconeogenesis, fatty acid synthesis and catabolism of fat and glucose, actions that are independent of AMPK could also be involved. Some of these actions are described in this chapter, and some have been discussed in reviews by Wang et al. (2014) and Kim et al. (2014).

5 Lowering of blood cholesterol, blood pressure and incidence of cardiovascular diseases

5.1 Studies in humans

The alleviation of MetS by tea logically leads to the reduction of the risks for CVDs (reviewed in Deka and Vita (2011), Di Castelnuovo et al. (2012) and Munir et al., (2013)). The lowering of plasma cholesterol levels and blood pressure as well as improvement of insulin sensitivity and endothelial function by green tea has been verified in many studies (reviewed in Munir et al. (2013)). In a review of 11 RCTs, both green and black teas decreased LDL cholesterol and blood pressure (Hartley et al., 2013). The strongest evidence for the reduction of CVD risk by the consumption of green tea was provided by large cohort studies in Japan. In the Ohsaki National Health Insurance Cohort Study (n = 40 530), deaths due to CVDs were decreased dose-dependently by tea consumption at quantities of one to more than five cups of tea per day (Kuriyama et al., 2006b). In another study with 76 979 Japanese adults, the consumption of green tea was also associated with decreased CVD mortality, but daily consumption of more than six cups of tea was needed to manifest the effect (Mineharu et al., 2011). Correlation between the consumption of tea and the decreased risk of stroke was reported by two studies from China and Japan (Liang et al., 2009; Kokubo et al., 2013). A meta-analysis of 14 prospective studies, covering 513 804 participants with a median follow-up of 11.5 years, found an inverse association between tea consumption and risk of stroke, and the protective effect of green tea appeared to be stronger than that of black tea (Shen et al., 2012). Many, but not all, studies in the United States and Europe demonstrated an inverse association between black tea consumption and CVD risk (Arab et al., 2009; de Koning Gans et al., 2010; Mukamal et al., 2006; Sesso et al., 2003; Deka and Vita, 2011). A meta-analysis, including 6 case–control and 12 cohort studies (5 measured green tea and 13 measured black tea as the exposure), found a reduced risk of coronary artery disease by 28% via green tea consumption (10% decrease in risk with an increment in consumption of one cup per day). However, there was no significant protective effect from black tea (Wang et al., 2011).

5.2 Possible mechanisms

As discussed above, reduction of body weight and alleviation of MetS by tea consumption would decrease the stress to the cardiovascular system. Beneficial effects of tea catechins in lowering plasma cholesterol levels, preventing hypertension and improving endothelial function contribute to the prevention of CVDs. The cholesterol-lowering effect is likely due to the decrease in cholesterol absorption or reabsorption by catechins as well as the decrease in cholesterol synthesis via the inhibition of 3-hydroxy-3-methylglutaryl-coenzyme A reductase (mediated by the activation of AMPK). Enhanced nitric oxide signalling has been suggested as a common mechanism for catechins to decrease blood pressure and the severity of myocardial infarction (Munir et al., 2013). Several studies showed that green tea or black tea polyphenols increased endothelial nitric oxide synthase (eNOS) activity in bovine aortic endothelial cells and rat aortic rings (Jochmann et al., 2008; Lorenz et al., 2004; Aggio et al., 2013), possibly mediated by AMPK and other signal pathways. Tea catechins may suppress the expression of caveolin-1, a negative regulator of eNOS (Li et al., 2009). Similarly, EGCG has also been shown to lower the expression of endothelin-1, possibly through the activation of AMPK, and this reduces vasoconstrictor tone and may directly increase bioavailability of nitric oxide to improve endothelial function (Akiyama et al., 2009). AMPK has also been suggested to regulate the antioxidant status (Colombo and Moncada, 2009) and to mediate the anti-inflammatory activity of EGCG in endothelial cells (Wu et al., 2014). EGCG has also been shown to induce the expression of haeme oxygenase 1 in aortic endothelial cells (Pullikotil et al., 2012), and this may increase anti-inflammatory activity to benefit the cardiovascular system. While moderate doses of EGCG have yielded beneficial effects, a very high dose (1% in diet) has been shown to promote, rather than to attenuate, vascular inflammation in hyperglycemic mice (Pae et al., 2012).

6 Neuroprotective effects of tea

Several epidemiological studies have suggested that tea drinking was associated with the improvement of cognitive function. For example, green tea consumption was associated with a lower prevalence of cognitive impairment among elderly Japanese (Kuriyama et al., 2006a; Tomata et al., 2012) and with better cognitive performance in community-living elderly Chinese (Feng et al., 2010). Other studies have described a moderate risk reduction of Parkinson's disease (PD) in tea drinkers (Barranco Quintana et al., 2009; Li et al., 2012). A recent meta-analysis of 13 articles involving 901 764 participants showed a linear dose relationship for decreased PD risk with tea, coffee and caffeine consumption (Qi and Li, 2014). Another meta-analysis of 20 observational epidemiological studies involving 31 479 participants, however, found that caffeine intake from tea or coffee was not associated with the risk of cognitive disorders (including dementia, Alzheimer's disease and cognitive impairment/decline) (Kim et al., 2015). A recent review article suggested that tea, coffee and caffeine may be protective against late-life cognitive impairment/decline, but the association was not found in all cognitive domains and lacked a distinct dose–response association (Panza et al., 2015). Tea drinking was also associated with lowering risks of depressive symptoms (Hintikka et al., 2005) and psychological distress (Hozawa et al., 2009).

Laboratory studies also showed that EGCG protected against neurodegenerative diseases such as Parkinson's and Alzheimer diseases in rodent models (Levites et al., 2001, Rezai-Zadeh et al., 2005). More recent studies have also demonstrated the neuroprotective effects of tea catechins against okadaic acid-induced acute learning and memory impairment in rate (Li et al., 2014) and against 6-hydroxydopamine-induced behavioural and depressive changes in a rat model of PD (Bitu Pinto et al., 2015). Interestingly, a recent study in an *N*-methyl-4-phenyl-1,2,3,6-terahydropyridine-induced Parkinsonian monkey model demonstrated that oral administration of tea polyphenols alleviates motor impairment, dopaminergic neuronal injury and cerebral α-synuclein aggregation (Chen et al., 2015). There are also studies suggesting beneficial effects of dietary EGCG in combination with exercise on brain health and slows the progression of Alzheimer's disease in TgCRNDS mice (Walker et al., 2015). Similarly, EC plus exercise are also neuroprotective against cognitive deficiencies and progression of moderate or mid-stage Alzheimer's disease in APP/PS1 mice (Zhang et al., 2016).

Based on animal and cell culture studies, catechins and theanine are also considered to be responsible for the neuroprotective action of green tea. The proposed mechanisms of action of EGCG and catechins include their antioxidant and iron-chelating activities as well as anti-inflammatory and signal-modulating activities related to neuronal cell growth (reviewed in Weinreb et al. (2009) and Mandel et al. (2008)). Since neuronal cells are sensitive to oxidative damage, antioxidant actions of catechins are important for the protective effect. Iron accumulation in brain is a common feature of neurodegeneration, and iron is also involved in the production of amyloid precursor protein and β-amyloid formation (Rogers et al., 2002). EGCG is known to reverse the iron-dependent events in many *in vitro* models. Although several reports indicated that EGCG can pass the blood–brain barrier and exert a neuroprotective action, it is unclear whether EGCG can reach the brain at levels high enough for neuroprotection in humans.

The characteristic amino acid theanine, which can cross the blood–brain barrier, is considered to be an important compound for the neuroprotective actions of tea. Several studies indicated that theanine relieved anxiety symptoms in patients with schizophrenic and schizoaffective disorders (Ritsner et al., 2011), and a combination of GTE and theanine improved memory and attention in subjects with mild cognitive impairments (Park et al., 2011). Theanine was also effective in improving sleep quality in boys diagnosed with attention-deficit hyperactivity disorder (Lyon et al., 2011). As an analogue of glutamate, a neuroexcitatory transmitter, theanine may act as an antagonist against glutamate receptors and prevent glutamate-induced excitatory neuronal toxicity. Since the affinity of theanine to the receptors is low, interference in glutamine transporter and modulation of levels of other neurotransmitters, such as dopamine and γ-aminobutyric acid, have also been suggested to be neuroprotective mechanisms of theanine (Kakuda, 2011).

Overall, there are many human and animal studies suggesting the protective effects of tea consumption against several types of cognitive dysfunctions. Longer follow-up studies on a large number of subjects are needed to fully confirm these effects.

7 Conclusion

As discussed above, many laboratory and epidemiological studies strongly suggest the beneficial health effects of tea consumption in the prevention of chronic diseases. Evidence from human intervention studies, however, is lacking concerning some of the beneficial effects, such as the prevention of cancer. In animal studies, conditions are usually optimized to demonstrate the hypothesized effects, and the doses of tea preparations used are usually higher than the levels of human consumption. Effects of interfering factors, such as smoking, physical activity and dietary intake of coffee, calories, fat and fibre, make it difficult to interpret human data. After taking these factors into consideration, a clearer pattern may emerge.

The possible molecular mechanisms by which tea constituents lower body weight, alleviate MetS and prevent diabetes, CVDs, neurodegenerative diseases and cancer are summarized in Fig. 2. The antioxidant activities of tea polyphenols are believed to contribute all these beneficial effects. These involve the direct antioxidant effect of polyphenols, their chelation of metal ions and their activation of the Nrf2-mediated cellular defence system. The binding of tea polyphenols to lipids and proteins (including inhibitions of digestive enzymes) and decreasing macronutrient absorption in the intestine appear to be a major mechanism for body weight reduction in animals and humans who

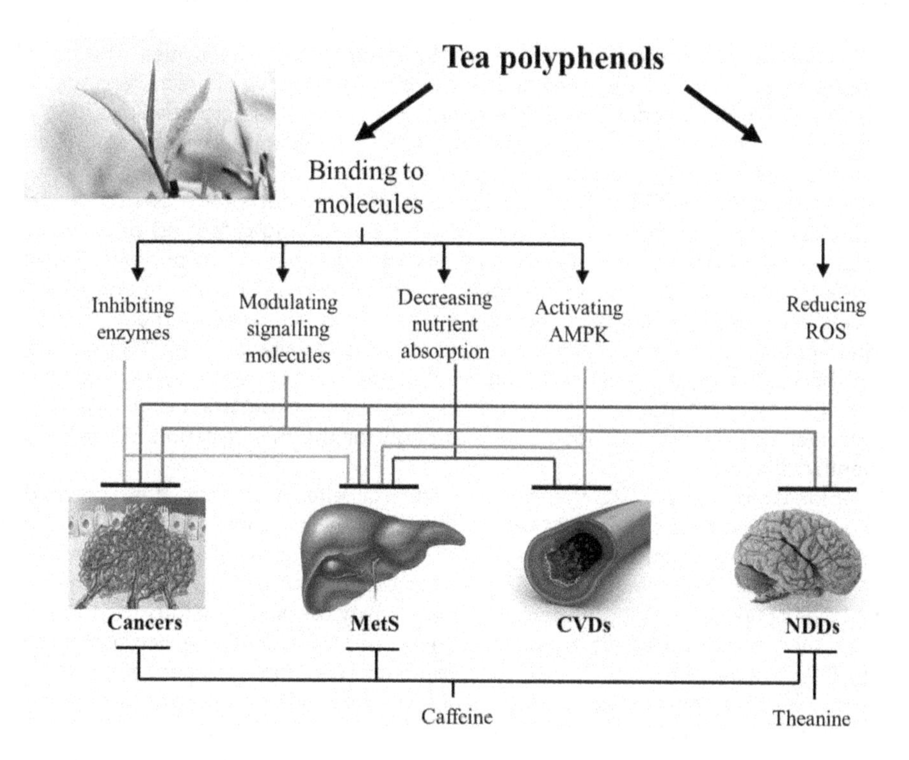

Figure 2 Possible mechanisms for the prevention of chronic disease by tea constituents. Abbreviation: NDDs = neurodegenerative diseases.

ingest excessive amount of calories. The body weight-lowering effect provides beneficial effects to alleviate or prevent MetS, diabetes and CVDs and perhaps also cancer. The activation of AMPK, possibly a consequence of polyphenols binding to mitochondrial election transport proteins, is also expected to alleviate MetS and diabetes, which reduces the risk of CVDs. For the prevention of neurodegenerative diseases, although tea polyphenols have been showing activity in laboratory studies, epidemiology studies suggest the important contribution of caffeine, because coffee has similar effects. Theanine could play a major role in preventing neurodegeneration and improving mental health. Multiple mechanisms have been proposed for the cancer-preventive activities of tea polyphenols and caffeine. The binding of catechins, such as EGCG, to specific enzymes, receptors and signalling molecules provides exciting insights for further investigations on the mechanisms of prevention of many of these diseases. How a molecule such as EGCG could specifically affect the functions of many proteins and the expression of many genes as discussed in this chapter is intriguing and requires further investigation.

Bioavailability is an important issue in determining the biological effects of tea polyphenols in internal organs. This factor could explain the results that many of the beneficial effects were observed with green tea but not with black tea. Tea polyphenols that are not absorbed into the blood, however, may exert their effects in the gastrointestinal tract, for example, in decreasing lipid absorption. This is probably why black tea is also effective in lowering body weight, body fat and cholesterol levels. The intestinal microbiota may degrade tea polyphenols, as has been shown for catechins and theaflavins (Li et al., 2000; Chen et al., 2012). Some of the metabolites may have interesting biological activities. The microbial degradation of black tea polyphenols has not been sufficiently characterized and more research is needed. The effects of tea consumption on intestinal microorganisms have been studied (Axling et al., 2012; Jin et al., 2012; Janssens et al., 2016), and more comprehensive research on the microbiota using newer approaches would provide more information.

In studies dealing with MetS and diabetes, beneficial effects have been observed in individuals consuming three or four cups of tea (600–900 mg catechins) daily. However, such cause–effect and dose–response relationship in the prevention of other diseases is not clear. Intervention studies on the prevention of cancer, CVDs and neurodegenerative diseases are difficult to conduct. Because of budget considerations, most intervention studies are limited to a length of a few or several years, usually with high-risk populations. A drawback of using a high-risk population is that the latent disease may have progressed to a stage that is reflective to the intervention agents. If a general population is used, the intervention period may not be long enough for the manifestation of the disease-preventive effects.

In epidemiological studies, the lack of beneficial effects of green tea consumption observed in some studies could be due to the relatively low quantities of tea consumed. For protection against certain diseases, the effects of lower levels of tea consumption (one to three cups per day) may be subtle. However, caution should be applied in the use of high doses of tea extracts for disease prevention. Many cases of hepatotoxicity due to the consumption of GTE-based dietary supplements have been reported (reviewed in Mazzanti et al. (2009) and Koyama et al. (2010)). Because of the hepatotoxicity concern, French and Spanish regulatory agencies suspended market authorization of a weight reduction product containing GTEs (Sarma et al., 2008). In addition, because of the strong binding activities of tea polyphenols to minerals and biomolecules, ingestion of large quantities of tea extracts may cause nutritional and other problems, even though such problems are not expected to occur due to regular tea beverage consumption (Yang and Hong, 2013).

In order to gain a better understanding of the health effects of tea consumption, more research is needed. Some suggestions are as follows:

1 More laboratory studies to elucidate the biochemical basis for the reported health effects of different types of tea. The relevance of these results to human health should be evaluated.
2 More large and long-term prospective studies with special attention paid to the quantity and types of tea consumption, smoking status, diet, physical activities, genetic polymorphism and other possible interfering factors.
3 More well-designed intervention studies, based on strong laboratory data and with adequate duration.

8 Where to look for further information

For general information about tea, look at other chapters of this book. For the effects of tea on health, please see references (Yang et al. 2016; Yang and Hong, 2013; Panza et al. 2015).

9 Acknowledgements

This work was supported by NIH grants CA120915, CA122474 and CA133021. We thank Ms. Vi P. Dan for her assistance in the preparation of this manuscript.

10 References

Aggio, A., Grassi, D., Onori, E., D'alessandro, A., Masedu, F., Valenti, M. and Ferri, C. 2013. Endothelium/nitric oxide mechanism mediates vasorelaxation and counteracts vasoconstriction induced by low concentration of flavanols. *Eur. J. Nutr.*, 52, 263–72.

Akiyama, S., Katsumata, S., Suzuki, K., Nakaya, Y., Ishimi, Y. and Uehara, M. 2009. Hypoglycemic and hypolipidemic effects of hesperidin and cyclodextrin-clathrated hesperetin in Goto-Kakizaki rats with type 2 diabetes. *Biosci. Biotechnol. Biochem.*, 73, 2779–82.

Arab, L., Liu, W. and Elashoff, D. 2009. Green and black tea consumption and risk of stroke: A meta-analysis. *Stroke*, 40, 1786–92.

Arnaud, M. J. 2011. Pharmacokinetics and metabolism of natural methylxanthines in animal and man. In Fredholm, B. B. (Ed.), *Methylxanthines, Handbook of Experimental Pharmacology 200.* Berlin Heidelberg: Springer-Verlag.

Axling, U., Olsson, C., Xu, J., Fernandez, C., Larsson, S., Strom, K., Ahrne, S., Holm, C., Molin, G. and Berger, K. 2012. Green tea powder and Lactobacillus plantarum affect gut microbiota, lipid metabolism and inflammation in high-fat fed C57BL/6J mice. *Nutr. Metab. (Lond)*, 9, 105.

Balentine, D. A., Wiseman, S. A. and Bouwens, L. C. 1997. The chemistry of tea flavonoids. *Crit. Rev. Food Sci. Nutr.*, 37, 693–704.

Banerjee, S., Ghoshal, S. and Porter, T. D. 2012. Phosphorylation of hepatic AMP-activated protein kinase and liver kinase B1 is increased after a single oral dose of green tea extract to mice. *Nutr Res*, 32, 985–90.

Barranco Quintana, J. L., Allam, M. F., Del Castillo, A. S. and Navajas, R. F. 2009. Parkinson's disease and tea: a quantitative review. *J. Am. Coll. Nutr.*, 28, 1–6.

Bettuzzi, S., Brausi, M., Rizzi, F., Castagnetti, G., Peracchia, G. and Corti, A. 2006. Chemoprevention of human prostate cancer by oral administration of green tea catechins in volunteers with high-grade prostate intraepithelial neoplasia: a preliminary report from a one-year proof-of-principle study. *Cancer Res.*, 66, 1234–40.

Bitu Pinto, N., Da Silva Alexandre, B., Neves, K. R., Silva, A. H., Leal, L. K. and Viana, G. S. 2015. Neuroprotective Properties of the Standardized Extract from Camellia sinensis (Green Tea) and Its Main Bioactive Components, Epicatechin and Epigallocatechin Gallate, in the 6-OHDA Model of Parkinson's Disease. *Evid. Based Complement Alternat. Med.*, 2015, 161092.

Bogdanski, P., Suliburska, J., Szulinska, M., Stepien, M., Pupek-Musialik, D. and Jablecka, A. 2012. Green tea extract reduces blood pressure, inflammatory biomarkers, and oxidative stress and improves parameters associated with insulin resistance in obese, hypertensive patients. *Nutr. Res.*, 32, 421–7.

Bose, M., Lambert, J. D., Ju, J., Reuhl, K. R., Shapses, S. A. and Yang, C. S. 2008. The major green tea polyphenol, (–)-epigallocatechin-3-gallate, inhibits obesity, metabolic syndrome, and fatty liver disease in high-fat-fed mice. *J. Nutr.*, 138, 1677–83.

Byun, J. K., Yoon, B. Y., Jhun, J. Y., Oh, H. J., Kim, E. K., Min, J. K. and Cho, M. L. 2014. Epigallocatechin-3-gallate ameliorates both obesity and autoinflammatory arthritis aggravated by obesity by altering the balance among CD4+ T-cell subsets. *Immunol. Lett.*, 157, 51–9.

Cani, P. D., Neyrinck, A. M., Fava, F., Knauf, C., Burcelin, R. G., Tuohy, K. M., Gibson, G. R. and Delzenne, N. M. 2007. Selective increases of bifidobacteria in gut microflora improve high-fat-diet-induced diabetes in mice through a mechanism associated with endotoxaemia. *Diabetologia*, 50, 2374–83.

Chang, C. S., Chang, Y. F., Liu, P. Y., Chen, C. Y., Tsai, Y. S. and Wu, C. H. 2012. Smoking, habitual tea drinking and metabolic syndrome in elderly men living in rural community: The Tianliao old people (TOP) study 02. *PloS One*, 7, e38874.

Chen, H., Hayek, S., Rivera Guzman, J., Gillitt, N. D., Ibrahim, S. A., Jobin, C. and Sang, S. 2012. The microbiota is essential for the generation of black tea theaflavins-derived metabolites. *PLoS One*, 7, e51001.

Chen, M., Wang, T., Yue, F., Li, X., Wang, P., Li, Y., Chan, P. and Yu, S. 2015. Tea polyphenols alleviate motor impairments, dopaminergic neuronal injury, and cerebral alpha-synuclein aggregation in MPTP-intoxicated parkinsonian monkeys. *Neuroscience*, 286, 383–92.

Chen, Y. K., Cheung, C., Reuhl, K. R., Liu, A. B., Lee, M. J., Lu, Y. P. and Yang, C. S. 2011. Effects of green tea polyphenol (-)-epigallocatechin-3-gallate on newly developed high-fat/Western-style diet-induced obesity and metabolic syndrome in mice. *J. Agric. Food Chem.*, 59, 11862–71.

Chow, H. H., Cai, Y., Hakim, I. A., Crowell, J. A., Shahi, F., Brooks, C. A., Dorr, R. T., Hara, Y. and Alberts, D. S. 2003. Pharmacokinetics and safety of green tea polyphenols after multiple-dose administration of epigallocatechin gallate and polyphenon E in healthy individuals. *Clin. Cancer Res.*, 9, 3312–9.

Chow, H. H. and Hakim, I. A. 2011. Pharmacokinetic and chemoprevention studies on tea in humans. *Pharmacol. Res.*, 64, 105–12.

Collins, Q. F., Liu, H. Y., Pi, J., Liu, Z., Quon, M. J. and Cao, W. 2007. Epigallocatechin-3-gallate (EGCG), a green tea polyphenol, suppresses hepatic gluconeogenesis through 5'-AMP-activated protein kinase. *J. Biol. Chem.*, 282, 30143–9.

Colombo, S. L. and Moncada, S. 2009. AMPKalpha1 regulates the antioxidant status of vascular endothelial cells. *Biochem. J.*, 421, 163–9.

Crew, K. D., Brown, P., Greenlee, H., Bevers, T. B., Arun, B., Hudis, C., Mcarthur, H. L., Chang, J., Rimawi, M., Vornik, L., Cornelison, T. L., Wang, A., Hibshoosh, H., Ahmed, A., Terry, M. B., Santella, R. M., Lippman, S. M. and Hershman, D. L. 2012. Phase IB randomized, double-blinded, placebo-controlled, dose escalation study of polyphenon E in women with hormone receptor-negative breast cancer. *Cancer Prev. Res. (Phila)*, 5, 1144–54.

De Koning Gans, J. M., Uiterwaal, C. S., Van Der Schouw, Y. T., Boer, J. M., Grobbee, D. E., Verschuren, W. M. and Beulens, J. W. 2010. Tea and coffee consumption and cardiovascular morbidity and mortality. *Arterioscler. Thromb. Vasc. Biol.*, 30, 1665–71.

Deka, A. and Vita, J. A. 2011. Tea and cardiovascular disease. *Pharmacol. Res.*, 64, 136–45.

Di Castelnuovo, A., Di Giuseppe, R., Iacoviello, L. and De Gaetano, G. 2012. Consumption of cocoa, tea and coffee and risk of cardiovascular disease. *Eur. J. Intern. Med.*, 23, 15–25.

Everard, A., Lazarevic, V., Derrien, M., Girard, M., Muccioli, G. G., Neyrinck, A. M., Possemiers, S., Van Holle, A., Francois, P., De Vos, W. M., Delzenne, N. M., Schrenzel, J. and Cani, P. D. 2011. Responses of gut microbiota and glucose and lipid metabolism to prebiotics in genetic obese and diet-induced leptin-resistant mice. *Diabetes*, 60, 2775–86.

Everard, A., Matamoros, S., Geurts, L., Delzenne, N. M. and Cani, P. D. 2014. Saccharomyces boulardii administration changes gut microbiota and reduces hepatic steatosis, low-grade inflammation, and fat mass in obese and type 2 diabetic db/db mice. *MBio*, 5, e01011–14.

Feng, L., Gwee, X., Kua, E. H. and Ng, T. P. 2010. Cognitive function and tea consumption in community dwelling older Chinese in Singapore. *J. Nutr. Health aging*, 14, 433–8.

Ford, E. S. 2005. Prevalence of the metabolic syndrome defined by the International Diabetes Federation among adults in the U.S. *Diabetes Care*, 28, 2745–9.

Friedrich, M., Petzke, K. J., Raederstorff, D., Wolfram, S. and Klaus, S. 2012. Acute effects of epigallocatechin gallate from green tea on oxidation and tissue incorporation of dietary lipids in mice fed a high-fat diet. *Int. J. Obes. (Lond)*, 36, 735–43.

Gao, Y. T., Mclaughlin, J. K., Blot, W. J., Ji, B. T., Dai, Q. and Fraumeni, J. F., Jr. 1994. Reduced risk of esophageal cancer associated with green tea consumption. *J. Natl Cancer Inst.*, 86, 855–8.

Guan, F., Liu, A. B., Li, G., Yang, Z., Sun, Y., Yang, C. S. and Ju, J. H. 2012. Deleterious Effects of High Concentrations of (-)-Epigallocatechin-3-Gallate and Atorvastatin in Mice with Colon Inflammation. *Nutr. and Cancer*, 64, 847–55.

Hardie, D. G. 2015. AMPK: positive and negative regulation, and its role in whole-body energy homeostasis. *Curr. Opin. Cell Biol.*, 33, 1–7.

Hardie, D. G., Ross, F. A. and Hawley, S. A. 2012. AMPK: A nutrient and energy sensor that maintains energy homeostasis. *Nat. Rev. Mol. Cell Biol.*, 13, 251–62.

Hartley, L., Flowers, N., Holmes, J., Clarke, A., Stranges, S., Hooper, L. and Rees, K. 2013. Green and black tea for the primary prevention of cardiovascular disease. *Cochrane Database Syst. Rev.*, 6, CD009934.

Hintikka, J., Tolmunen, T., Honkalampi, K., Haatainen, K., Koivumaa-Honkanen, H., Tanskanen, A. and Viinamaki, H. 2005. Daily tea drinking is associated with a low level of depressive symptoms in the Finnish general population. *Eur. J. Epidemiol.*, 20, 359–63.

Hodgson, A. B., Randell, R. K., Boon, N., Garczarek, U., Mela, D. J., Jeukendrup, A. E. and Jacobs, D. M. 2013. Metabolic response to green tea extract during rest and moderate-intensity exercise. *J. Nutr. Biochem.*, 24, 325–34.

Hou, Z., Sang, S., You, H., Lee, M. J., Hong, J., Chin, K. V. and Yang, C. S. 2005. Mechanism of action of (-)-epigallocatechin-3-gallate: Auto-oxidation-dependent inactivation of epidermal growth factor receptor and direct effects on growth inhibition in human esophageal cancer KYSE 150 cells. *Cancer Res.*, 65, 8049–56.

Hozawa, A., Kuriyama, S., Nakaya, N., Ohmori-Matsuda, K., Kakizaki, M., Sone, T., Nagai, M., Sugawara, Y., Nitta, A., Tomata, Y., Niu, K. and Tsuji, I. 2009. Green tea consumption is associated with lower psychological distress in a general population: The Ohsaki Cohort 2006 Study. *Am. J. Clin. Nutr.*, 90, 1390–6.

Huang, J., Wang, Y., Xie, Z., Zhou, Y., Zhang, Y. and Wan, X. 2014. The anti-obesity effects of green tea in human intervention and basic molecular studies. *Eur. J. Clin. Nutr.*, 68, 1075–87.

Hursel, R., Viechtbauer, W., Dulloo, A. G., Tremblay, A., Tappy, L., Rumpler, W. and Westerterp-Plantenga, M. S. 2011. The effects of catechin rich teas and caffeine on energy expenditure and fat oxidation: A meta-analysis. *Obesity Rev.*, 12, e573–81.

Hursel, R., Viechtbauer, W. and Westerterp-Plantenga, M. S. 2009. The effects of green tea on weight loss and weight maintenance: a meta-analysis. *Int. J. Obes.*, 33, 956–61.

Huxley, R., Lee, C. M., Barzi, F., Timmermeister, L., Czernichow, S., Perkovic, V., Grobbee, D. E., Batty, D. and Woodward, M. 2009. Coffee, decaffeinated coffee, and tea consumption in relation to incident type 2 diabetes mellitus: A systematic review with meta-analysis. *Arch. Intern. Med.*, 169, 2053–63.

Iso, H., Date, C., Wakai, K., Fukui, M. and Tamakoshi, A. 2006. The relationship between green tea and total caffeine intake and risk for self-reported type 2 diabetes among Japanese adults. *Ann. Intern. Med.*, 144, 554–62.

James, K. D., Forester, S. C. and Lambert, J. D. 2015. Dietary pretreatment with green tea polyphenol, (-)-epigallocatechin-3-gallate reduces the bioavailability and hepatotoxicity of subsequent oral bolus doses of (-)-epigallocatechin-3-gallate. *Food Chem. Toxicol.*, 76, 103–8.

Janssens, P. L., Hursel, R. and Westerterp-Plantenga, M. S. 2015. Long-term green tea extract supplementation does not affect fat absorption, resting energy expenditure, and body composition in adults. *J. Nutr.*, 145, 864–70.

Janssens, P. L., Penders, J., Hursel, R., Budding, A. E., Savelkoul, P. H. and Westerterp-Plantenga, M. S. 2016. Long-Term Green Tea Supplementation Does Not Change the Human Gut Microbiota. *PLoS One*, 11, e0153134.

Jemnitz, K., Heredi-Szabo, K., Janossy, J., Ioja, E., Vereczkey, L. and Krajcsi, P. 2010. ABCC2/Abcc2: A multispecific transporter with dominant excretory functions. *Drug Metab. Rev.*, 42, 402–36.

Jin, J. S., Touyama, M., Hisada, T. and Benno, Y. 2012. Effects of green tea consumption on human fecal microbiota with special reference to Bifidobacterium species. *Microbiol. Immunol.*, 56, 729–39.

Jochmann, N., Lorenz, M., Krosigk, A., Martus, P., Bohm, V., Baumann, G., Stangl, K. and Stangl, V. 2008. The efficacy of black tea in ameliorating endothelial function is equivalent to that of green tea. *Br. J. Nutr.*, 99, 863–8.

Joe, A. K., Schnoll-Sussman, F., Bresalier, R. S., Abrams, J. A., Hibshoosh, H., Cheung, K., Friedman, R. A., Yang, C. S., Milne, G. L., Liu, D. D., Lee, J. J., Abdul, K., Bigg, M., Foreman, J., Su, T., Wang, X., Ahmed, A., Neugut, A. I., Akpa, E., Lippman, S. M., Perloff, M., Brown, P. H. and Lightdale, C. J. 2015. Phase Ib Randomized, Double-Blinded, Placebo-Controlled, Dose Escalation Study of Polyphenon E in Patients with Barrett's Esophagus. *Cancer Prev. Res. (Phila)*, 8, 1131–7.

Kakuda, T. 2011. Neuroprotective effects of theanine and its preventive effects on cognitive dysfunction. *Pharmacol. Res.*, 64, 162–8.

Kim, H. S., Quon, M. J. and Kim, J. A. 2014. New insights into the mechanisms of polyphenols beyond antioxidant properties; lessons from the green tea polyphenol, epigallocatechin 3-gallate. *Redox Biol.*, 2, 187–95.

Kim, S., Lee, M. J., Hong, J., Li, C., Smith, T. J., Yang, G. Y., Seril, D. N. and Yang, C. S. 2000. Plasma and tissue levels of tea catechins in rats and mice during chronic consumption of green tea polyphenols. *Nutr. Cancer*, 37, 41–8.

Kim, Y. S., Kwak, S. M. and Myung, S. K. 2015. Caffeine intake from coffee or tea and cognitive disorders: a meta-analysis of observational studies. *Neuroepidemiology*, 44, 51–63.

Kokubo, Y., Iso, H., Saito, I., Yamagishi, K., Yatsuya, H., Ishihara, J., Inoue, M. and Tsugane, S. 2013. The impact of green tea and coffee consumption on the reduced risk of stroke incidence in Japanese population: The Japan public health center-based study cohort. *Stroke*, 44, 1369–74.

Koyama, Y., Kuriyama, S., Aida, J., Sone, T., Nakaya, N., Ohmori-Matsuda, K., Hozawa, A. and Tsuji, I. 2010. Association between green tea consumption and tooth loss: Cross-sectional results from the Ohsaki Cohort 2006 Study. *Prev. Med.*, 50, 173–9.

Kumar, N. B., Pow-Sang, J., Egan, K. M., Spiess, P. E., Dickinson, S., Salup, R., Helal, M., Mclarty, J. W., Williams, C. R., Schreiber, F. M. O., Parnes, H. L., Sebti, S., Kazi, A., Kang, L., Quinn, G. P., Smith, T., Yue, B., Diaz, K., Chornokur, G., Crocker, T. and Schell, M. J. 2015. Randomized, Placebo-Controlled Trial of Green Tea Catechins for Prostate Cancer Prevention. *Cancer Prev. Res. (Phila)*. 8, 879–87.

Kuriyama, S., Hozawa, A., Ohmori, K., Shimazu, T., Matsui, T., Ebihara, S., Awata, S., Nagatomi, R., Arai, H. and Tsuji, I. 2006a. Green tea consumption and cognitive function: A cross-sectional study from the Tsurugaya Project 1. *Am. J. Clin. Nutr.*, 83, 355–61.

Kuriyama, S., Shimazu, T., Ohmori, K., Kikuchi, N., Nakaya, N., Nishino, Y., Tsubono, Y. and Tsuji, I. 2006b. Green tea consumption and mortality due to cardiovascular disease, cancer, and all causes in Japan: The Ohsaki study. *Jama*, 296, 1255–65.

Lee, L. S., Choi, J. H., Sung, M. J., Hur, J. Y., Hur, H. J., Park, J. D., Kim, Y. C., Gu, E. J., Min, B. and Kim, H. J. 2015. Green tea changes serum and liver metabolomic profiles in mice with high-fat diet-induced obesity. *Mol. Nutr. Food Res.*, 59, 784–94.

Lee, M.-J., Maliakal, P., Chen, L., Meng, X., Bondoc, F. Y., Prabhu, S., Lambert, G., Mohr, S. and Yang, C. S. 2002. Pharmacokinetics of tea catechins after ingestion of green tea and (-)-epigallocatechin-3-gallate by humans: Formation of different metabolites and individual variability. *Cancer Epidemiol. Biomark. Prev.*, 11, 1025–32.

Levites, Y., Weinreb, O., Maor, G., Youdim, M. B. and Mandel, S. 2001. Green tea polyphenol (-)-epigallocatechin-3-gallate prevents N-methyl-4-phenyl-1,2,3,6-tetrahydropyridine-induced dopaminergic neurodegeneration. *J. Neurochem.*, 78, 1073–82.

Li, C., Lee, M. J., Sheng, S., Meng, X., Prabhu, S., Winnik, B., Huang, B., Chung, J. Y., Yan, S., Ho, C. T. and Yang, C. S. 2000. Structural identification of two metabolites of catechins and their kinetics in human urine and blood after tea ingestion. *Chem. Res. Toxicol.*, 13, 177–84.

Li, F. J., Ji, H. F. and Shen, L. 2012. A meta-analysis of tea drinking and risk of Parkinson's disease. *Scientific World J.*, 2012, 923464.

Li, G. X., Chen, Y. K., Hou, Z., Xiao, H., Jin, H., Lu, G., Lee, M. J., Liu, B., Guan, F., Yang, Z., Yu, A. and Yang, C. S. 2010. Pro-oxidative activities and dose-response relationship of (-)-epigallocatechin-3-gallate in the inhibition of lung cancer cell growth: a comparative study in vivo and in vitro. *Carcinogenesis*, 31, 902–10.

Li, H., Wu, X., Wu, Q., Gong, D., Shi, M., Guan, L., Zhang, J., Liu, J., Yuan, B., Han, G. and Zou, Y. 2014. Green tea polyphenols protect against okadaic acid-induced acute learning and memory impairments in rats. *Nutrition*, 30, 337–42.

Li, N., Sun, Z., Han, C. and Chen, J. 1999. The chemopreventive effects of tea on human oral precancerous mucosa lesions. *Proc. Soc. Exp. Biol. Med.*, 220, 218–24.

Li, Y., Ying, C., Zuo, X., Yi, H., Yi, W., Meng, Y., Ikeda, K., Ye, X., Yamori, Y. and Sun, X. 2009. Green tea polyphenols down-regulate caveolin-1 expression via ERK1/2 and p38MAPK in endothelial cells. *J. Nutr. Biochem.*, 20, 1021–7.

Liang, W., Lee, A. H., Binns, C. W., Huang, R., Hu, D. and Zhou, Q. 2009. Tea consumption and ischemic stroke risk: A case-control study in southern China. *Stroke*, 40, 2480–5.

Lipinski, C. A., Lombardo, F., Dominy, B. W. and Feeney, P. J. 2001. Experimental and computational approaches to estimate solubility and permeability in drug discovery and development settings. *Adv. Drug Deliv. Rev.*, 46, 3–26.

Long, Y. C. and Zierath, J. R. 2006. AMP-activated protein kinase signaling in metabolic regulation. *J. Clin. Invest.*, 116, 1776–83.

Lorenz, M., Wessler, S., Follmann, E., Michaelis, W., Dusterhoft, T., Baumann, G., Stangl, K. and Stangl, V. 2004. A constituent of green tea, epigallocatechin-3-gallate, activates endothelial nitric oxide synthase by a phosphatidylinositol-3-OH-kinase-, cAMP-dependent protein kinase-, and Akt-dependent pathway and leads to endothelial-dependent vasorelaxation. *J. Biol. Chem.*, 279, 6190–5.

Lyon, M. R., Kapoor, M. P. and Juneja, L. R. 2011. The effects of L-theanine (Suntheanine(R)) on objective sleep quality in boys with attention deficit hyperactivity disorder (ADHD): A randomized, double-blind, placebo-controlled clinical trial. *Altern. Med. Rev.*, 16, 348–54.

Mandel, S. A., Amit, T., Kalfon, L., Reznichenko, L. and Youdim, M. B. 2008. Targeting multiple neurodegenerative diseases etiologies with multimodal-acting green tea catechins. *J. Nutr.*, 138, 1578S–1583S.

Mazzanti, G., Menniti-Ippolito, F., Moro, P. A., Cassetti, F., Raschetti, R., Santuccio, C. and Mastrangelo, S. 2009. Hepatotoxicity from green tea: A review of the literature and two unpublished cases. *Eur. J. Clin. Pharmacol.*, 65, 331–41.

Mielgo-Ayuso, J., Barrenechea, L., Alcorta, P., Larrarte, E., Margareto, J. and Labayen, I. 2014. Effects of dietary supplementation with epigallocatechin-3-gallate on weight loss, energy homeostasis, cardiometabolic risk factors and liver function in obese women: Randomised, double-blind, placebo-controlled clinical trial. *Br. J. Nutr.*, 111, 1263–71.

Mineharu, Y., Koizumi, A., Wada, Y., Iso, H., Watanabe, Y., Date, C., Yamamoto, A., Kikuchi, S., Inaba, Y., Toyoshima, H., Kondo, T. and Tamakoshi, A. 2011. Coffee, green tea, black tea and oolong tea consumption and risk of mortality from cardiovascular disease in Japanese men and women. *J. Epidemiol. Community Health*, 65, 230–40.

Mukamal, K. J., Alert, M., Maclure, M., Muller, J. E. and Mittleman, M. A. 2006. Tea consumption and infarct-related ventricular arrhythmias: The determinants of myocardial infarction onset study. *J. Am. Coll. Nutr.*, 25, 472–9.

Mulder, T. P., Van Platerink, C. J., Wijnand Schuyl, P. J. and Van Amelsvoort, J. M. 2001. Analysis of theaflavins in biological fluids using liquid chromatography-electrospray mass spectrometry. *J. Chromatogr. B Biomed. Sci. Appl.*, 760, 271–9.

Munir, K. M., Chandrasekaran, S., Gao, F. and Quon, M. J. 2013. Mechanisms for food polyphenols to ameliorate insulin resistance and endothelial dysfunction: Therapeutic implications for diabetes and its cardiovascular complications. *Am. J. Physiol. Endocrinol. Metab.*, 305, E679–86.

Murase, T., Misawa, K., Haramizu, S. and Hase, T. 2009. Catechin-induced activation of the LKB1/AMP-activated protein kinase pathway. *Biochem. Pharmacol.*, 78, 78–84.

Ogunleye, A. A., Xue, F. and Michels, K. B. 2010. Green tea consumption and breast cancer risk or recurrence: A meta-analysis. *Breast Cancer Res. Treat.*, 119, 477–84.

Okuda, M. H., Zemdegs, J. C., De Santana, A. A., Santamarina, A. B., Moreno, M. F., Hachul, A. C., Dos Santos, B., Do Nascimento, C. M., Ribeiro, E. B. and Oyama, L. M. 2014. Green tea extract improves high fat diet-induced hypothalamic inflammation, without affecting the serotoninergic system. *J. Nutr. Biochem.*, 25, 1084–9.

Ortsater, H., Grankvist, N., Wolfram, S., Kuehn, N. and Sjoholm, A. 2012. Diet supplementation with green tea extract epigallocatechin gallate prevents progression to glucose intolerance in db/db mice. *Nutr. Metab. (Lond)*, 9, 11.

Pae, M., Ren, Z., Meydani, M., Shang, F., Smith, D., Meydani, S. N. and Wu, D. 2012. Dietary supplementation with high dose of epigallocatechin-3-gallate promotes inflammatory response in mice. *J. Nutr. Biochem.*, 23, 526–31.

Panza, F., Solfrizzi, V., Barulli, M. R., Bonfiglio, C., Guerra, V., Osella, A., Seripa, D., Sabba, C., Pilotto, A. and Logroscino, G. 2015. Coffee, tea, and caffeine consumption and prevention of late-life cognitive decline and dementia: a systematic review. *J. Nutr. Health Aging*, 19, 313–28.

Park, S. K., Jung, I. C., Lee, W. K., Lee, Y. S., Park, H. K., Go, H. J., Kim, K., Lim, N. K., Hong, J. T., Ly, S. Y. and Rho, S. S. 2011. A combination of green tea extract and l-theanine improves memory and attention in subjects with mild cognitive impairment: A double-blind placebo-controlled study. *J. Med. Food*, 14, 334–43.

Pham, N. M., Nanri, A., Kochi, T., Kuwahara, K., Tsuruoka, H., Kurotani, K., Akter, S., Kabe, I., Sato, M., Hayabuchi, H. and Mizoue, T. 2014. Coffee and green tea consumption is associated with insulin resistance in Japanese adults. *Metabolism*, 63, 400–8.

Phung, O. J., Baker, W. L., Matthews, L. J., Lanosa, M., Thorne, A. and Coleman, C. I. 2010. Effect of green tea catechins with or without caffeine on anthropometric measures: A systematic review and meta-analysis. *J. Clin. Nutr.*, 91, 73–81.

Pullikotil, P., Chen, H., Muniyappa, R., Greenberg, C. C., Yang, S., Reiter, C. E., Lee, J. W., Chung, J. H. and Quon, M. J. 2012. Epigallocatechin gallate induces expression of heme oxygenase-1 in endothelial cells via p38 MAPK and Nrf-2 that suppresses proinflammatory actions of TNF-alpha. *J. Nutr. Biochem.*, 23, 1134–45.

Qi, H. and Li, S. 2014. Dose-response meta-analysis on coffee, tea and caffeine consumption with risk of Parkinson's disease. *Geriatr. Gerontol. Int.*, 14, 430–9.

Qin, B., Polansky, M. M., Harry, D. and Anderson, R. A. 2010. Green tea polyphenols improve cardiac muscle mRNA and protein levels of signal pathways related to insulin and lipid metabolism and inflammation in insulin-resistant rats. *Mol. Nutr. Food Res.*, 54 Suppl 1, S14–23.

Rezai-Zadeh, K., Shytle, D., Sun, N., Mori, T., Hou, H., Jeanniton, D., Ehrhart, J., Townsend, K., Zeng, J., Morgan, D., Hardy, J., Town, T. and Tan, J. 2005. Green tea epigallocatechin-3-gallate (EGCG) modulates amyloid precursor protein cleavage and reduces cerebral amyloidosis in Alzheimer transgenic mice. *J. Neurosci.*, 25, 8807–14.

Ritsner, M. S., Miodownik, C., Ratner, Y., Shleifer, T., Mar, M., Pintov, L. and Lerner, V. 2011. L-theanine relieves positive, activation, and anxiety symptoms in patients with schizophrenia and schizoaffective disorder: An 8-week, randomized, double-blind, placebo-controlled, 2-center study. *J. Clin. Psychiatry*, 72, 34–42.

Rogers, J. T., Randall, J. D., Cahill, C. M., Eder, P. S., Huang, X., Gunshin, H., Leiter, L., Mcphee, J., Sarang, S. S., Utsuki, T., Greig, N. H., Lahiri, D. K., Tanzi, R. E., Bush, A. I., Giordano, T. and Gullans, S. R. 2002. An iron-responsive element type II in the 5'-untranslated region of the Alzheimer's amyloid precursor protein transcript. *J. Biol. Chem.*, 277, 45518–28.

Sae-Tan, S., Grove, K. A. and Lambert, J. D. 2011. Weight control and prevention of metabolic syndrome by green tea. *Pharmacol. Res.*, 64, 146–54.

Sang, S., Lambert, J. D., Ho, C. T. and Yang, C. S. 2011. The chemistry and biotransformation of tea constituents. *Pharmacol. Res.*, 64, 87–99.

Sarma, D. N., Barrett, M. L., Chavez, M. L., Gardiner, P., Ko, R., Mahady, G. B., Marles, R. J., Pellicore, L. S., Giancaspro, G. I. and Low Dog, T. 2008. Safety of green tea extracts: a systematic review by the US Pharmacopeia. *Drug Saf.*, 31, 469–84.

Sasazuki, S., Tamakoshi, A., Matsuo, K., Ito, H., Wakai, K., Nagata, C., Mizoue, T., Tanaka, K., Tsuji, I., Inoue, M. and Tsugane, S. 2012. Green tea consumption and gastric cancer risk: an evaluation based on a systematic review of epidemiologic evidence among the Japanese population. *Jpn. J. Clin. Oncol.*, 42, 335–46.

Scheid, L., Ellinger, S., Alteheld, B., Herholz, H., Ellinger, J., Henn, T., Helfrich, H. P. and Stehle, P. 2012. Kinetics of L-theanine uptake and metabolism in healthy participants are comparable after ingestion of L-theanine via capsules and green tea. *J. Nutr.*, 142, 2091–6.

Serisier, S., Leray, V., Poudroux, W., Magot, T., Ouguerram, K. and Nguyen, P. 2008. Effects of green tea on insulin sensitivity, lipid profile and expression of PPARalpha and PPARgamma and their target genes in obese dogs. *Br. J. Nutr.*, 99, 1208–16.

Serrano, J. C., Gonzalo-Benito, H., Jove, M., Fourcade, S., Cassanye, A., Boada, J., Delgado, M. A., Espinel, A. E., Pamplona, R. and Portero-OTIN, M. 2013. Dietary intake of green tea polyphenols regulates insulin sensitivity with an increase in AMP-activated protein kinase alpha content and changes in mitochondrial respiratory complexes. *Mol. Nutr. Food Res.*, 57, 459–70.

Sesso, H. D., Gaziano, J. M., Liu, S. and Buring, J. E. 2003. Flavonoid intake and the risk of cardiovascular disease in women. *Am. J. Clin. Nutr.*, 77, 1400–8.

Shen, G., Xu, C., Hu, R., Jain, M. R., Nair, S., Lin, W., Yang, C. S., Chan, J. Y. and Kong, A. N. 2005. Comparison of (-)-epigallocatechin-3-gallate elicited liver and small intestine gene expression profiles between C57BL/6J mice and C57BL/6J/Nrf2 (-/-) mice. *Pharm. Res.*, 22, 1805–20.

Shen, L., Song, L. G., Ma, H., Jin, C. N., Wang, J. A. and Xiang, M. X. 2012. Tea consumption and risk of stroke: A dose-response meta-analysis of prospective studies. *J. Zhejiang Univ. Sci. B*, 13, 652–62.

Song, Y., Manson, J. E., Buring, J. E., Sesso, H. D. and Liu, S. 2005. Associations of dietary flavonoids with risk of type 2 diabetes, and markers of insulin resistance and systemic inflammation in women: A prospective study and cross-sectional analysis. *J. Am Coll. Nutr.*, 24, 376–84.

Suliburska, J., Bogdanski, P., Szulinska, M., Stepien, M., Pupek-Musialik, D. and Jablecka, A. 2012. Effects of green tea supplementation on elements, total antioxidants, lipids, and glucose values in the serum of obese patients. *Biol. Trace Elem. Res.*, 149, 315–22.

Sun, C. L., Yuan, J. M., Koh, W. P., Lee, H. P. and Yu, M. C. 2007. Green tea and black tea consumption in relation to colorectal cancer risk: The Singapore Chinese Health Study. *Carcinogenesis*, 28, 2143–8.

Sun, C. L., Yuan, J. M., Koh, W. P. and Yu, M. C. 2006. Green tea, black tea and colorectal cancer risk: A meta-analysis of epidemiologic studies. *Carcinogenesis*, 27, 1301–9.

Takami, H., Nakamoto, M., Uemura, H., Katsuura, S., Yamaguchi, M., Hiyoshi, M., Sawachika, F., Juta, T. and Arisawa, K. 2013. Inverse correlation between coffee consumption and prevalence of metabolic syndrome: Baseline survey of the Japan Multi-Institutional Collaborative Cohort (J-MICC) Study in Tokushima, Japan. *J. Epidemiol.*, 23, 12–20.

Tao, L., Forester, S. C. and Lambert, J. D. 2014. The role of the mitochondrial oxidative stress in the cytotoxic effects of the green tea catechin, (-)-epigallocatechin-3-gallate, in oral cells. *Mol. Nutr. Food Res.*, 58, 665–76.

Tomata, Y., Kakizaki, M., Nakaya, N., Tsuboya, T., Sone, T., Kuriyama, S., Hozawa, A. and Tsuji, I. 2012. Green tea consumption and the risk of incident functional disability in elderly Japanese: The Ohsaki Cohort 2006 Study. *Am. J. Clin. Nutr.*, 95, 732–9.

Tsao, A. S., Liu, D., Martin, J., Tang, X. M., Lee, J. J., El-Naggar, A. K., Wistuba, I., Culotta, K. S., Mao, L., Gillenwater, A., Sagesaka, Y. M., Hong, W. K. and Papadimitrakopoulou, V. 2009. Phase II randomized, placebo-controlled trial of green tea extract in patients with high-risk oral premalignant lesions. *Cancer Prev. Res. (Phila)*, 2, 931–41.

Valenti, D., De Bari, L., Manente, G. A., Rossi, L., Mutti, L., Moro, L. and Vacca, R. A. 2013. Negative modulation of mitochondrial oxidative phosphorylation by epigallocatechin-3 gallate leads to growth arrest and apoptosis in human malignant pleural mesothelioma cells. *Biochim. Biophys. Acta*, 1832, 2085–96.

Van Der Pijl, P. C., Chen, L. and Mulder, T. P. J. 2010. Human disposition of L-theanine in tea or aqueous solution. *J. Functional Food*, 2, 239–44.

Vernarelli, J. A. and Lambert, J. D. 2013. Tea consumption is inversely associated with weight status and other markers for metabolic syndrome in US adults. *Eur. J. Nutr.*, 52, 1039–48.

Walker, J. M., Klakotskaia, D., Ajit, D., Weisman, G. A., Wood, W. G., Sun, G. Y., Serfozo, P., Simonyi, A. and Schachtman, T. R. 2015. Beneficial effects of dietary EGCG and voluntary exercise on behavior in an Alzheimer's disease mouse model. *J. Alzheimers Dis.*, 44, 561–72.

Wang, D., Wang, Y., Wan, X., Yang, C. S. and Zhang, J. 2015. Green tea polyphenol (-)-epigallocatechin-3-gallate triggered hepatotoxicity in mice: Responses of major antioxidant enzymes and the Nrf2 rescue pathway. *Toxicol. Appl. Pharmacol.*, 283, 65–74.

Wang, H., Wen, Y., Du, Y., Yan, X., Guo, H., Rycroft, J. A., Boon, N., Kovacs, E. M. and Mela, D. J. 2010. Effects of catechin enriched green tea on body composition. *Obesity*, 18, 773–9.

Wang, J., Zhang, W., Sun, L., Yu, H., Ni, Q. X., Risch, H. A. and Gao, Y. T. 2012. Green tea drinking and risk of pancreatic cancer: A large-scale, population-based case-control study in urban Shanghai. *Cancer epidemiol.*, 36, e354–e358.

Wang, S., Moustaid-Moussa, N., Chen, L., Mo, H., Shastri, A., Su, R., Bapat, P., Kwun, I. and Shen, C. L. 2014. Novel insights of dietary polyphenols and obesity. *J. Nutr. Biochem.*, 25, 1–18.

Wang, Z. M., Zhou, B., Wang, Y. S., Gong, Q. Y., Wang, Q. M., Yan, J. J., Gao, W. and Wang, L. S. 2011. Black and green tea consumption and the risk of coronary artery disease: a meta-analysis. *J. Clin. Nutr.*, 93, 506–15.

Weinreb, O., Amit, T., Mandel, S. and Youdim, M. B. 2009. Neuroprotective molecular mechanisms of (-)-epigallocatechin-3-gallate: a reflective outcome of its antioxidant, iron chelating and neuritogenic properties. *Genes Nutr.*, 4, 283–96.

Wu, A. H., Spicer, D., Stanczyk, F. Z., Tseng, C. C., Yang, C. S. and Pike, M. C. 2012. Effect of 2-month controlled green tea intervention on lipoprotein cholesterol, glucose, and hormone levels in healthy postmenopausal women. *Cancer Prev. Res. (Phila)*, 5, 393–402.

Wu, J., Xu, X., Li, Y., Kou, J., Huang, F., Liu, B. and Liu, K. 2014. Quercetin, luteolin and epigallocatechin gallate alleviate TXNIP and NLRP3-mediated inflammation and apoptosis with regulation of AMPK in endothelial cells. *Eur. J. Pharmacol.*, 745, 59–68.

Yamashita, Y., Wang, L., Tinshun, Z., Nakamura, T. and Ashida, H. 2012. Fermented tea improves glucose intolerance in mice by enhancing translocation of glucose transporter 4 in skeletal muscle. *J. Agric. Food Chem.*, 60, 11366–71.

Yamashita, Y., Wang, L., Wang, L., Tanaka, Y., Zhang, T. and Ashida, H. 2014. Oolong, black and pu-erh tea suppresses adiposity in mice via activation of AMP-activated protein kinase. *Food Funct.*, 5, 2420–9.

Yang, C. S. and Hong, J. 2013. Prevention of chronic diseases by tea: possible mechanisms and human relevance. *Annu. Rev. Nutr.*, 33, 161–81.

Yang, C. S., Sang, S., Lambert, J. D. and Lee, M. J. 2008a. Bioavailability issues in studying the health effects of plant polyphenolic compounds. *Mol. Nutr. and Food Res.*, 52 Suppl 1, S139–51.

Yang, C. S., Sang, S., Lambert, J. D. and Lee, M. J. 2008b. Bioavailability issues in studying the health effects of plant polyphenolic compounds. *Mol. Nutr. Food Res.*, 52 Suppl 1, S139–51.

Yang, C. S., Wang, H., Li, G. X., Yang, Z., Guan, F. and Jin, H. 2011a. Cancer prevention by tea: Evidence from laboratory studies. *Pharmacol. Res.*, 64, 113–22.

Yang, C. S., Wang, X., Lu, G. and Picinich, S. C. 2009. Cancer prevention by tea: Animal studies, molecular mechanisms and human relevance. *Nat. Rev. Cancer.*, 9, 429–39.

Yang, C. S., Zhang, J., Zhang, L., Huang, J. and Wang, Y. 2016. Mechanisms of body weight reduction and metabolic syndrome alleviation by tea. *Mol. Nutr. Food. Res.*, 60, 160–74.

Yang, G., Shu, X. O., Li, H., Chow, W. H., Ji, B. T., Zhang, X., Gao, Y. T. and Zheng, W. 2007. Prospective cohort study of green tea consumption and colorectal cancer risk in women. *Cancer Epidemiol. Biomarkers Prev.*, 16, 1219–23.

Yang, G., Zheng, W., Xiang, Y. B., Gao, J., Li, H. L., Zhang, X., Gao, Y. T. and Shu, X. O. 2011b. Green tea consumption and colorectal cancer risk: A report from the Shanghai Men's Health Study. *Carcinogenesis*, 32, 1684–8.

Yuan, J. M., Sun, C. and Butler, L. M. 2011. Tea and cancer prevention: epidemiological studies. *Pharmacol. Res.*, 64, 123–35.

Zhang, Z., Wu, H. and Huang, H. 2016. Epicatechin Plus Treadmill Exercise are Neuroprotective Against Moderate-stage Amyloid Precursor Protein/Presenilin 1 Mice. *Pharmacogn. Mag.*, 12, S139–46.

Zhou, J., Farah, B. L., Sinha, R. A., Wu, Y., Singh, B. K., Bay, B. H., Yang, C. S. and Yen, P. M. 2014. Epigallocatechin-3-gallate (EGCG), a green tea polyphenol, stimulates hepatic autophagy and lipid clearance. *PLoS One*, 9, e87161.

CPSIA information can be obtained
at www.ICGtesting.com
Printed in the USA
JSHW061456090623
42991JS00002B/44

9 781786 769756